James M. McCanney, M.S.

PLANET-X COMETS & EARTH CHANGES

A SCIENTIFIC TREATISE ON THE EFFECTS OF A NEW LARGE PLANET OR COMET ARRIVING IN OUR SOLAR SYSTEM AND EXPECTED EARTH WEATHER AND EARTH CHANGES

Published By

jmccanneyscience.com press

Minneapolis, Minnesota

6th Printing - 2003

ISBN 0-9722186-0-2

PLANET-X
COMETS
& EARTH CHANGES

About the author and this text:

James M. McCanney, M.S. received a sound classical physics training at St. Mary's University, receiving a Bachelor of Arts degree with a double major in physics and mathematics in 1970. He was offered full scholarship awards to three major US physics graduate schools to pursue his graduate physics studies. However, he chose instead to postpone graduate studies for a period of three years while he traveled and taught Physics and Mathematics in Spanish in Latin America.

During this time he spent a good deal of time traveling to ruins of ancient cities and archeological sites, studying firsthand many times as the ruins were dug from under dirt that had not been moved for thousands of years. Also during this time he developed the basis for his theoretical work that would, at a later date, deal with the celestial mechanics of N-bodies and plasma physics. It was here also that he learned to appreciate the fact that the ruins and devastation he was witnessing had to have come from celestial events that were so devastating that they left the Earth and these stone cities in ruins, in some cases, leaving no trace of the inhabitants.

With this new understanding of archeology, astronomy of the ancients, physics and the world around him, Mr. McCanney returned to graduate school in 1973 and earned a master's degree in nuclear and solid-state physics from Tulane University, New Orleans, LA. He was again offered a full fellowship to continue on with Ph.D. studies, but once again he declined and returned to Latin America to study archeology and teach physics, mathematics and computer science in Spanish. He continued his work to explore the mysteries of celestial mechanics and its relationship to the planets, moons and other celestial bodies.

In 1979 he joined the faculty of Cornell University, Ithaca N.Y. as an introductory instructor in physics. It was during this time that he had access to NASA data returning daily from the Voyager I and II spacecraft as they traveled by the planets Jupiter, Saturn and beyond (as well as data from many other space craft). It was here he recognized that his theoretical work regarding the electro-dynamic nature of the solar system and universe had its signatures in the new data that was streaming in from the edges of the solar system. All standard science continued to look at gravitational explanations for the working of the planets, moons and other objects of the solar system, while Mr. McCanney was applying his electro-dynamic scientific theories, and ventured to say for the first time that comets were not dirty snowballs. His papers were published at first in the standard astrophysical journals, but soon he began to receive resistance from the standard astronomical community and within a short period of time, the journals would no longer publish his theoretical work. Mr. McCanney was removed from his teaching position because of his beliefs regarding the electro-dynamic nature of the solar system.

Contrary to the traditional belief that the solar system formed all at one time 4.5 billion years ago and has not changed significantly since, Mr. McCanney's theoretical work essentially stated that the solar system was dynamic and adopting new members on an ongoing basis. He pointed to the planet Venus, the Jovian moon Io, the Saturnian moon Titan and the small planet Pluto (which supports an atmosphere even though it is so distant from the warmth of the Sun and has insufficient gravity to hold an atmosphere for long) as being obvious new members of our solar system. He stated that all this was proof that the way this occurred was by "planetary capture".

His theoretical work additionally stated that comets were not dirty snowballs, but were large electrical "vacuum cleaners" in outer space. The comets were drawing in vast amounts of material by way of powerful electrical forces and there was potential for very large comets capable of disrupting the planetary structure that was already in place. His innovative theories on plasma physics and a new model for fusion in the solar atmosphere provided the basis for the electric fields and plasma discharge phenomena that have become the core elements of his theoretical models of the true nature of the solar system in which we live.

Upon being fired from the physics department for his radical beliefs, Mr. McCanney was rehired shortly thereafter by the mathematics department also at Cornell University, where he taught for another year and a half and continued to publish his papers in astrophysical journals. Once again astronomers forced his removal and he was once again blackballed from publishing in the astrophysics journals in1981.

During this time Mr. McCanney established himself as the originator of the theoretical work regarding the electrical nature of the cosmos, which today is being proven correct on an ongoing basis by space-probes returning data from outer space. Many of his predictions such as x-rays to the sunward side of comet nuclei, that comet nuclei would be found to have no ice or water frozen on their surfaces and that comets interact electrically with the Sun to affect Earth weather, have now been confirmed by direct measurements in 1986, 1996, 2001 and 2002 respectively. Many other more abstract concepts have also been verified.

There exists a rare combination of factors that makes Mr. McCanney a unique person who stands alone in the development of the scientific theories summarized in this book. Some have tried to borrow and copy this work, but when observers consider the factors involved they too will agree that the extensive rewriting of standard scientific structures had to be accomplished by someone with a rare set of characteristics and circumstances.

He was always at the top of his classes in Mathematics and Physics and was always creating his own formulas and proofs. His education was soundly based in classical and modern physics. He was able to recognize that when the basic new aspects of the functioning of the solar system were understood and then verified in space probe data, he had the ability to extend this information and take it to all its logical conclusions. This all occurred while working in and around the top rated scientists of the day at Cornell University who were still at least 2 decades behind what Mr. McCanney was discovering and writing.

Another unique condition was that Cornell University offered a rare location since it was not only a Library of Congress (if it was in print it was there), but also it was a repository of data for NASA. Armed with his existing theoretical work and this incredible source of information, and with the timing that coincided with the daily arrival of new data from the Voyager and other space craft from the far reaches of the solar system, he was in a totally unique position to do what he has done.

An essential requirement of anyone who attempts to alter the fundamental propositions of a subject as complex as astronomy and astrophysics, is an in depth knowledge of the history of that and all related sciences. Mr. McCanney has studied the history of science extensively and understands where the theories came from that currently make up the structures of science. He has the rare ability to speak at many levels. He has taken a topic as complex as plasma physics and celestial mechanics (the study of rarified gasses and the motion of the celestial bodies) and simplify it in terms that the general public can understand. That is why his university classes always grew rapidly. He does not make an ego trip out of trying to impress his audiences with big words and unusual terms as do many university professors or the recent influx of internet hoaxsters.

There are few people who would have the tenacity to pursue and uphold their beliefs for as long as he has had to do in facing the odds pitted against him over the past decades, and to emerge in tact with as full a commitment as when he started that long ago. His personal philosophy states that he will not accept government funding on the principle that it has corrupted the scientific process.

These numerous and individually rare characteristics make the record clear that the important contributions made here combine both personal traits and a situation of "being in the right place at the right time" as the space craft data poured into Cornell University as Mr McCanney's theoretical ideas were solidifying.

In 1981 the Interdisciplinary Journal KRONOS agreed to publish what has since become known in inner circles as the "3 Part comet paper", which is republished in it's entirety at the end of this book along with other historically significant publications of Mr. McCanney. His work today includes many new significant insights into the connection between the Sun, comets, earth weather, The Sun-Earth Connection and Earth Changes. The results of that knowledge will become obvious when you read this book.

Mr. McCanney has also remained active and well-known within the space science - astronomy community and within professional societies, and although standard astronomers still resist accepting his theoretical work, he is generally well respected amongst his piers in these communities when attending professional conferences. He is what some have called "the last of the independent scientists" who were able to work "on the inside" and still remain active to talk about it on the "outside".

In the mid-1990s Mr. McCanney's work was recognized by a group of high-level Russian scientists who had measured but did not understand electro-dynamic effects around Earth and in the solar system. They translated all of his papers to date into Russian. These are being taught at the university level as the leading edge of research in this field. It is only due to the ongoing and intentional efforts of NASA that his work has received such little attention in the western scientific community and press.

It should be pointed out also that no new or bizarre fundamental laws of physics are needed to understand the electro-dynamic nature of the universe. This entire textbook is based on standard physics and simply introduces the principles of the working of our solar system based on a new understanding of how the Sun produces fusion and therefore electric fields (and magnetic fields) in the solar system. Once these fundamental principles are put in place, the rest is simply the logical conclusion of standard laws of physics being applied to the situation.

It is very interesting to note that historical records of the major disasters witnessed by the ancients are now understood in light of this new understanding of how our solar system works. A scientific basis and the "theoretical key" has now been introduced to give fundamental understanding of the Biblical references regarding the End Days, The accounts of the Exodus from the Legends of the Jews, the Mayan accounts of Quetzalcoatl (the plumed serpent god of the night sky whose heart became the planet Venus), the accounts of the Egyptian mythological planet Nibiru, the accounts and historical records and writings of the destruction of Atlantis as well as the writings of such world renowned authors as Casey, Sitchin, Velikovsky, Blavatsky, Donnelly and others who could see beyond the standard and incorrect explanations of where we came from and the true history of man and our planet.

This book is a summary of information that spans over 25 years of theoretical effort and is needed on the international market to inform the public of the true nature of our past relative to celestial events, and additionally, of the dangers that stand before us as we deal with the possibility of new planetary members being adopted into the solar system. It is also an integral addition to a series of books that is being authored by a unique group of scholars. These are serious authors who have "paid their dues" and whose words have been solidified over decades of research and painful attention to detail. This book is an essential addition to that series that already includes treatises on symbolism, calendar changes caused by man's need to adopt to celestial events and our connections to the past through our correct interpretation of history.

This text does not make any specific predictions of dates of such an arrival nor does it make any predictions or analysis regarding any particular planetary object(s) that might come into the solar system. It's purpose is to discuss the many possible solar system changes and Earth Changes that we might experience if and when such an object does come into the solar system.

This book also attempts to give for posterity a current view of the "politics of science" of the time we live in as well as to dispel many of the myths being propagated by government supported scientists who still adhere to incorrect outdated theories. It additionally dispels misinformation generated from the growth of Internet quacks, some of whom are attempting to plagiarize and sell this information, but who have no certification or training in this complicated area of science.

a.s.

TABLE OF CONTENTS

PREFACE

Putting this book into perspective

When the land mass that is now called the United States of America was first explored in the 1800's, every naturalist that passed through this new land talked of the evidence … the entire land had been ravaged by a Great Flood. Alaska was no exception. This led scholars such as Ignatious Donnelly (then Lieutenant Governor of the territory of Minnesota) to write his book "Atlantis the Antediluvian World". Raised in the coal fields of Pennsylvania, Donnelly was keenly aware of the geophysical evidence that Earth had seen many radical changes that shaped its morphology.

It has been noted by many, that in the quest to tame this rugged land, man had destroyed most if not all of the remnants of the Great Flood by the year 1900. So today it is easy for some to ignore the words of those forefathers who saw the massive destruction with their own eyes, and to claim now that the Great Flood and these remnants never existed, or to minimize it as some smaller localized event.

When Aristotle's uncle Solon traveled to Egypt, his intention was to research for the young Greek nation the history of the human race. Still in existance were the ancient priest cults who explained to him that there was a great worldwide cataclysm that devastated the entire Earth. They told of a massive war that was ongoing between the pre-catastrophe Greek ancestors and a powerful technologically advanced nation called Atlantis, whose population spanned the globe. They stated that the Greeks were beginning to actually win in this war when the catastrophe hit ***all of Earth*** and the mighty nation of Atlantis sunk under the ocean in a single night. It was clearly more than just a localized flood or a few days of heavy rain. The entire Earth was shaken to its knees.

They additionally noted that the survivors that generated the new Greek nation were but simple sheep herders who survived the devastation and Great Flood, living in caves in the mountains of what we now call Greece. The Egyptian priest cults explained that their records went back 10,000 years and that every time their society got up and running, they were crushed anew by one of these major events.

These were not an average pestilence or famine, but a worldwide devastation that sent them back to the cave ages.

Today we are finding the remnants of these ancient societies. There were huge cities in what we now call South America long before the so-called Ice Age or the alleged crossing of the Bering Strait by early hunting tribes. What we are learning is that the history that is being taught as "standard knowledge" is totally wrong. There is a tremendous effort in numerous sectors of modern society to keep the truth of our real past from being told. The seats of knowledge in our universities, certain religious centers and the governments of western society are doing their very best to prevent this information from reaching the general public. They fear that this knowledge will set people free from the domination they are currently under.

If this is your first introduction to this subject, I can only in tell you with certitude that this is one of the most complicated and far-reaching topics that you will ever encounter. Thousands of extremely talented scholars have spent their entire careers researching in and around the subjects that are affected by the topics covered in this book. It affects every aspect of every discipline from astronomy and planetary geology to paleontology and biological evolution, from history of man and politics to the origins of religion.

The "key" that this book offers is a sound scientific framework that gives a firm direction to understand our past, present and future. It is backed by a person who "has been there" and can stand up to the onslaughts of rooms full if angry resentful astronomers and pompous NASA space scientists. It is not a "fluffy" book that simply repeats over and over with different catch phrases that a given theory explains everything you ever wanted to know, but with no substance. This book gets into the details of how the solar system works and why you will see and experience certain phenomena … the "Earth Changes". It also explains so the average person will fully understand not only the basic theory and principles, but the ramifications of those basic concepts as the causes of Earth weather and Earth Changes. It also clarifies some basic misconceptions propagated by well-intentioned people who do not have a scientific background (e.g. the causes of pole shifts and effects of magnetic fields, etc).

This book is but a minimal summary of the very extensive work by the current author and is not meant to be viewed as an all-inclusive work. Many individual topics could produce entire treatises on just those topics. This book is published mainly to describe to the average person the simplified results of very complicated plasma physics and celestial mechanics. It can also be viewed by scientists as an introductory outline on which to build a new structure and view of the solar system if they have the courage to shed the shroud of traditional astronomical theory.

The author has presented this material on what might be viewed as ***three levels***. One is an introductory general level such as is contained in this book, the second level includes a more complete outline of the physical principles involved in the new view of how our solar system functions (such as in the reprinted papers found in the Appendices of this book) , and the third level would be appropriate for astrophysicists, astronomers and space scientists with Ph.D.'s.

An example of the third level presentation was a lecture given by the author at Los Alamos National Laboratories in 1986 when this information was first presented to approximately 50 of the country's top plasma physicists in an invited talk entitled "Are comets really dirty snowballs". It was clear that this theoretical information was new to this physics community and also that it was extremely complicated to the extent that the full ramifications of this theory were not appreciated by even this very high level group of scientists.

This is now obvious to the average person as scientists continued to incorrectly design many space craft and continued to ignore and misunderstand the electrical nature of the cosmos (e.g. a good example was in allowing a 2 mile long tether to be extended out of the space shuttle which caused an enormous electrical discharge that resulted, nearly destroying the spacecraft and crew).

This author has spent over 20 years battling with great difficulty the scientific community that is centered around NASA and the American university system of scientists, as well as the scientific journal publication system and the "top down" government funding system of scientific research. This will be seen throughout this text to be a fundamental problem in getting the correct information to the general public regarding how outer space works in reality, and the possible effects it could have on your daily lives.

Now we are at the juncture where today as I write this information, NASA is frantically attempting to copy as much of this material as they can to pretend that they are presenting it to the public by a introducing what they call "space weather" on nightly television weather programs. They are putting out a text book to counter this book called "Storms from the Sun" which I encourage you to read and compare to this book (although they will never encourage you to read this book). When you read both books you will realize what 15 billion dollars annually has produced from NASA, and what I have produced in my own mind with no supercomputers, no funding and a with just a pencil and piece of paper.

Another difference between the two books is the method of presenting. NASA's book is written by professional "science writers" who interface with a selected team of "experts". The attitude is that the public is not intelligent enough to understand how things work, so they will just grace you with the results. My book assumes that the general public is quite intelligent and it explains how things work as well as the results so you can make your own decisions. I have laid my cards out on the table and have not taken the route of giving only bland peripheral statements so I could change or alter them at a later date if I were wrong, as does the NASA book.

As individual theoretical predictions of my work have come to pass and have been recognized in satellite data coming from outer space, NASA has repeatedly pretended that they have discovered this "radical" new information. Examples of this are my predictions of the "sweeper moons" found in the rings of Saturn (in this case the NASA scientist even stole the terminology from my published papers), the electrical nature of the outer planets, x-rays found to the sunward side of comets, that comet nuclei have no snow or water or ice anywhere on their surfaces (and additionally are not out-gassing any water or other gasses), but are found to be hot dry rocks, and the list goes on.

The unfortunate part of all of this thievery, is that they still can't get it right. They continue to mix these fundamentally new ideas with their old outdated and incorrect ideas to form a bigger mess than what they had before. I call this the "creeping crud". This is the continual altering of explanations by NASA to accommodate the totally unexpected results of space probe data that contradict their predictions. In the case of my theoretical work, read the papers at the end of this book to see that after 25 years they remain unaltered and true to the data that has been collected. In fact, NASA's explanations are daily "creeping" to become almost the same as mine, except they are having a hard time crossing the final frontier to say that I am right about the true nature of comets and that the solar system is not static. If they did, there is a huge house of cards around them that would come tumbling down (which might include a small item called congressional funding).

I unfortunately must dwell on this essential understanding of who and what NASA is for a number of paragraphs and later in a chapter dedicated to this subject. As this topic gains more and more public attention, NASA will work harder and harder to put their name and their incorrect and borrowed information (mixed with their outdated and incorrect theories) in the public eye through news releases, TV and weather specials and newspaper articles. This is already happening as this book goes to press.

It is essential that the public recognize them for who they are and what they are. It is also essential that the public understand the correct information so they're are not lead down the wrong path by what appear to be well educated scientists who stand behind their Ph.D.'s and government funding, not unlike the Wizard of Oz manipulating behind a very thin cloak to hide the reality that is NASA.

It is also NASA's job to prevent public panic, so they will always attempt to tell people that they are on top of the space science weather game and not to worry. Their apparent unity is not due to correct theoretical basis, but because anyone who disagrees within their ranks would be eliminated. It will be up to the educated public to decide who is correct as they witness and experience "Earth Changes" if we are unfortunate enough to have a new large comet and planet interact with our Sun and Earth.

The question is not whether or not it will happen, the only question is "when". To this question I take a quote from a famous historical figure who was asked this question when speaking of the End Days and celestial intervention … "of that day no man knows the time or place". Even if a large new object were known about today with exact location and orbital information, its orbit will change on a daily basis and the true orbit and location and time will elude prediction. So making predictions regarding WHEN new planets, comets and Earth Changes might take place is literally impossible although rough estimates can be made if an actual candidate is identified.

Thus, there is a requirement for this book today and in the future to give a solid explanation to the general public on the workings of the solar system. It is also essential for the public to be able to decide for themselves how to predict and interpret the myriad of damaging effects that could come to pass should a new large object arrive in our solar system. This information will also be based on past events and now gives a much better idea of the meaning of prophecies and predictions that come from the ancient texts and Scriptures. The ancients understood the severity of these events, and tried to pass this on to future generations.

The reader may initially want to read this book out of order. I know that many readers would first be interested in knowing of the predictions of the Earth Changes and solar system changes, that is, the results of the more complicated scientific explanations. If this is the case, I would recommend reading the chapters dealing with Earth Changes and the chapter that deals with pole shifts, mass extinctions, etc. Return then to read the other chapters that deal with the more scientific explanations of how the solar system works, to assure that the reader has a sound understanding of why these results come to be.

The Appendices contain for the readers benefit the original papers published by the this author dealing with the scientific theories on the new nature of the solar system. Once again this is only the basis of work that has proceeded for the past 25 years. The papers in the appendices were published (some almost 20 years ago and based on the original books published in the late 1970's), but these still hold true today as much as they did at that time. In fact part of the reason for including them here (duplicated exactly in their original form) is to illustrate how far ahead of the scientific community this author stood when those papers were originally published.

One last word of advise regarding those who will attempt to defray the impact of this book. It is already known that the established scientific community is positioning itself to counter this book and the success I have had in communicating these concepts to the public through international radio and other media. Each person will have to decide for themselves who is telling the truth (and who has an ulterior motive for attacking this work). Once again it is best to take heed from the greatest source of all. My response to them is taken from the same great man who spoke on a mound regarding the "End Day Prophesies", ... he said, "the learned and recognized scholars of the time will attempt to lead you astray". And when asked how the common person will know who to believe, that man answered simply, ... "you will know the truth when you hear it, for the truth is written in every man's heart".

I. Planet X, Comets and Earth Changes

News Flash! Comets are not Dirty Snowballs!

And an Introduction to: The Plasma Discharge Comet Model

If there were any phrase that must ring out loud and clear and dominate the pages of this book it is the above statement that "comets are not dirty snowballs"! With this basic realization, the standard theories for how the solar system came to be and how it works undergo the domino principle (one topples, then the next and the next until all have fallen).

In their place will go an entirely new set of principles and concepts that are essential to understanding how the solar system formed, but more importantly, that new objects (including some that are large and will become new planets and moons) may be adopted at any time.

In 1997, I attended a national scientific meeting in Baltimore, MD., whose main focus was to discuss the discovery of x-rays in Comet Hyakutake and the properties of the huge new comet Hale-Bopp. I had a large color poster made that had the above expression **(News Flash! Comets are not Dirty Snowballs)** written in large bold red letters at the top right corner. It should have been the nail in the coffin for the beleaguered and sinking "dirty snowball comet model". Let's get serious … x-rays coming from a docile little snow ball?!

Additionally my theoretical work had distinctly predicted that such x-rays would be found to the sunward side of comets almost 20 years prior to this experimental verification. When I got to the meeting, all the speaker slots were filled with NASA scientists (including one young Japanese fellow who could hardly speak English on his first time out). The news release meetings were all scheduled in secret so I would not be allowed to talk to the press when they arrived to cover the scientific conference results.

One of the leading NASA scientists had taken the paper I had submitted for publication and essentially plagiarized it, but mixed it of course with the dirty snow ball comet model resulting in a bizarre and unbelievably inaccurate picture of how comets work by either NASA standards or my own. But of course the drone news agencies printed the "results" without asking an intelligent question or batting an eye.

When the European Space Agency's Giotto space-craft arrived at Comet Halley in 1986, it was the only vehicle of the small fleet of craft sent to study that comet that had an on board camera. For an interesting account of that event I have included an article written after that fly by entitled "Fluff, Fluffier and Fluffiest". In this note of sarcasm I trust you will see what has happened all too many times as data from space probes have arrived from the far reaches of the solar system that blatantly contradicted the expectations of highly paid elite NASA scientists. Someday I may write a book that covers just that topic since there would be ample material to fill a volume or two

<u>A "GRIM" FAIRY TALE IN THREE ACTS</u>

FLUFF, FLUFFIER, & FLUFFIEST

(OR- How to constantly revise reality to protect the accepted theories *at all costs*)

by James McCanney

Act I - FLUFF

It happened in an instant...
it was 1986 and the European Giotto space craft (the only satellite going to comet Halley that had an onboard camera) was approaching that famous comet's nucleus through the clouds of dust and gas that surrounded it. As the available sunlight faded due to the dense clouds eclipsing the solar light, the only light came from the comet nucleus still buried deep in the cloud. Clue#1 ... why is the comet nucleus illuminated when the sunlight is blocked by the dense comet coma?

Hmmmm, could this be a "self luminous" object with an energy source *other* than sunlight (e.g. the plasma discharge current striking the comet nucleus)? and clue#2 ... how can the comet continue to form a tail from "solar radiation sublimating ices off the comet nucleus" when the sunlight is so completely blocked by the dense comet coma? The alleged process seems to be at best self-defeating! No one at NASA or ESO or Harvard or Yale or Cornell or JPL or anywhere else ... not even the news media ... thinks to ask such difficult questions as they are all beaming with "the right stuff".

On with our story ... an aging Fred Whipple (originator of the dirty snow ball comet model) has been invited to witness the "confirmation" of the dirty snow ball comet model. All the calculations had been done, all the journal articles had been refereed and all the Ph.D.s had been given to the bright new crop of comet scientists at all the best universities after years of hard study. The sales of books written by NASA scientists were brisk and the public had purchased and read and applauded TV specials that aired days before the famous fly-by of this famous comet.

A comet was said to be a sparkling white snowy ball with some dust that was released as the sun's heat burned off the outer layers. Artists' full color artwork dramatically reproduced the scientists' predictions and expectations. The Giotto space craft moved closer to the comet nucleus sending back picture after picture ... each one better defining the pure white potato shaped "snow ball".

Excitement rose to fever pitch as small "jets" could now be seen, and the pure white nucleus became better defined with each returning photo. At last, as the pure white nucleus, that Whipple and all the other scientists knew had to lie in the center of all the dust and gas, became clearly defined, Whipple screams out on international television ...**'it's the nucleus.. there it is... it's the nucleus!!!!'**

Just then the Giotto space craft broke through the final layers of gas and dust to reveal what is now known to be the darkest, blackest object ever photographed by any space craft in the history of the space program. Silence and shock filled the direct feed television and Giotto's picture broke up.

Something terrible had gone wrong. There was no snowy nucleus. What was it? And how could all the refereed journal articles have been wrong? The nucleus was not a loosely packed dirty snowball, but was a pitted burnt carbonaceous chondrite (rock). Dead silence filled the room. Seconds later the Giotto space craft started to gyrate wildly and the last signals received from it arrived with the electron particle counter off scale, circuits in overload and the magnetic sensors off the scale.

Surely the space craft had been designed to exceed all expected criteria by the comet specialists at ESO and NASA. An intense electron beam had hit Giotto. Giotto was temporarily rendered inoperable as it passed directly to the sunward side of comet Halley's nucleus. But from where could such an electron beam come from? The dirty snow ball theory never said anything about anything like that.

Act II - FLUFFIER

The theorists worked feverishly throughout the night to recalculate in an attempt to make the theory fit the data. No attempt was made to look at alternative comet theories, and no one dared mention that the cherished dirty snowball model might be incorrect.

One scientist postulated that if the "snow and ices" were in deep pits on the surface of the nucleus then we wouldn't see them but the solar radiation might penetrate and eek them out. Yeahhh! Everyone exclaimed and by morning the slightly dirtier than expected (but still dirty) "dirty snow ball model" was reconstructed. The Science News media specialist quickly constructed "apologetics articles" stating that the nucleus was a bit dirtier than expected, but none-the-less the dirty snowball model had been "confirmed", of course by the proponents who had no opposition since they had fired and destroyed the careers of anyone who had objected over the past 20 years.

NOW, let's look at how far this "theory" had to be stretched to make things fit. Prior "calculations" never took into account that there would be so little solar radiation at the comet nucleus, so right here we are dealing with a factor of thousands of times less energy than the original calculations needed to make the snow ball model work in the first place.

And now they were confronted with the slight (but never insurmountable) problem that the "snow and ice" was not even visible on the nucleus of the comet, and at most could not have been located over more than 1 to 2 percent of the surface area. The sunlight now had to "burrow in and ferret out" the ice in these hidden pockets. With less light energy than a Christmas tree bulb at a distance of 100 feet, they had to somehow show that tons of ices were sublimating per second. Additionally, certain areas had enough force to blast dust out with such force that the jets would be visible from Earth a mere 100 million miles away. Never mind that pesky high-energy electron beam that magically appeared, they had their hands full with just reconstructing the dirtier than expected nucleus.

Other features of the photos should have told them something also. For example the light was coming from the nucleus and shining outwards getting fainter as one looked away from the nucleus. If this light were from the Sun (the only source of energy in their model), then the light should have been brighter to the sun ward side of the nucleus, and dimming as it reached the nucleus. As before no one ever bothered to notice or ask those questions. Actually some were asking those questions but they were labeled as trouble- makers, "pseudo scientists" and "crackpots" and anyway, who would listen to someone who had been fired for incompetence and trying to publish "incorrect science".

Part III - FLUFFIEST

We are a full 14 years since the Halley's debacle. Other space probes have now confirmed that there is no water or ice on a comet nucleus and additionally that there are no cracks or pockets of ice either. Scientists are now hoping that the ice and snow is somewhere beneath the hot rocky surfaces they have recorded.

But even if there were water somewhere locked inside, there is no **WATER PRODUCTION** that would have to be there to account for tons of water and ice moving out from the nucleus per second. Essentially, the nucleus would have to look like a massive steam locomotive, but instead, when a comet nucleus is observed at close range, no water or ice or out-gassing of any kind is observed. In brief, the Dirty snowball comet model has been disproven by those who have Ph.D.s in this and they simply refuse to give up the ghost.

In addition to these "crimes" against science, other data were taken of Comet Halley including spectrographic data collected by professional astronomers with professional grade equipment (paid for with your tax dollars) that showed the comet nucleus giving off a distinctive red shift. That data was squelched too. Comets certainly do not give off red shifted light, a property usually only associated with massive stars and galaxies (the cause of this red shift is detailed in the appendix of my previously published works).

Not more than 6 years later a new comet entered the solar system. It had been here before not more than 4200 years ago. We called it Hale-Bopp. The ancients certainly had another name for it, as they did for the planet Venus, which was a huge comet before it was a planet in the not too distant past. But of course, the ancients who observed these and other comets, and handed down the legends of the damage they inflicted on planet Earth, were ill trained savages with no knowledge of orbital mechanics. Their fanciful tales of Jupiter throwing lightning bolts across the heavens to Mars and the many "Earth Changes" were just figments of their primitive imaginations according to NASA.

With a budget of 14 billion dollars a year that is congress proof, and with the largest collection of equipment and Ph.D.s ever amassed by mankind, the elite core of scientists at NASA could not possibly be wrong. Comets by edict are dirty snow balls and anyone who suggests otherwise must be a poorly educated ridiculously misguided pseudo scientist with a secret agenda to gain notoriety amongst the populous by creating fear and misinformation about what all good followers know are harmless tiny little snowballs.

And everyone lived happily ever after.

James McCanney
(author of The Plasma Discharge Comet Model)

In September of 2001, NASA's Deep Space 1 spacecraft photographed the nucleus of comet Borrelly. Although the comet was a well-developed comet with a full tail, the nucleus showed absolutely no water production, no ice, no snow ... ***no water of any kind emanating from the comet nucleus***. Essentially they had disproven the dirty snowball comet model. The news release, however, containing once again an apologetic response from NASA scientists (stating that there must be water *inside* Borrelly's nucleus), did not receive a press release until ***April 2002***. They did not address the obvious and irrefutable fact that *the nucleus was not **producing** any water* ... as mentioned, it should have looked like a run-away steam locomotive to produce the amount of water visible in Borrely's well developed tail. NASA's statement suggesting there might be water inside the tarred black nucleus was totally irrelevant. There is unquestionably a pattern of illegal insider trading occurring at NASA. This is the epitome of government funded science gone awry.

Many years before, a comet named Shoemacher Levy 9 crashed into the far side of the planet Jupiter. NASA's published prediction was that by the time Jupiter rotated for us to see the impact sites, the minuscule little pieces of dirty snow would have buried themselves far into the atmosphere of Jupiter and there would be nothing to look at. To the contrary, the Russians measured the after effects of explosions so massive that the x-rays and ultraviolet light reflecting off the moons of Jupiter could be detected at planet Earth. The Russians estimated that the energy expended in each explosion far exceeded all of the nuclear armaments of all countries on Earth exploding at one time. The rumble in the scientific community was that certainly the dirty snowball comet model could not possibly account for this.

What was not so obvious to most people was a subtle and almost unnoticed property of the seven tiny cometary pieces as they entered the realm of Jupiter. While approaching in the region outside of the realm of Jupiter, the comets were quite normal water comets. As soon as the comet nuclei entered the immediate area around Jupiter, the tails magically became completely and totally composed of $SO2$ (sulfur dioxide, with no water detected). This is of course is due to the fact that the comet derives its tail material from the local environment as is explained in the Plasma Discharge Comet Model. NASA scientists made no effort at explaining this and of course could not since it makes absolutely no sense when viewed with the dirty snowball model.

The list of contradictions goes on and on and on. At this point it is safe to say that the NASA scientists are in complete denial. They hide behind a news release system in which no one can ask them questions, they hide behind their referee journal system in which they referee their own articles, or as in the case of automatic publication in SCIENCE and NATURE magazines, articles are not refereed at all. They are scamming the public now, but history will look back on them with 20-20 hindsight. As these scientists read this book, and I know they will, I hope that each asks himself or herself if they truly want to be remembered this way, or will they choose the right path and go public and proclaim what they now know to be true. There's nothing wrong with being wrong. However, there is something wrong when they finally realize that this is the case, and maintain their staunch image like the unbending oak tree that resists the wind until it finally cracks completely. My own experience shows that the general public does not believe NASA anymore.

More than just the confirmation of a new way of viewing the workings of the solar system, what is far more important is that for every day that NASA sets back, says nothing and collects an ill gotten pay check, it is but another day the human civilization on this planet goes unprepared for a critical time of survival as a species. Each scientist must look within himself or herself and decide how much longer they are willing to collect a paycheck doing what they know is wrong. If a few scientists would break rank, the floodgates would open and the rush to move in the direction outlined in this book would become the dominant theme of the scientific community. My prediction (based on NASA's past performance), however, is that many will follow the lead of a few in power at these agencies and will attempt an all out theft of this material. Luckily, 25 years spans the time that this material has been in circulation, and once again history will see any such attempts for what it is.

Introduction to: The Plasma Discharge Comet Model (The true nature of comets and the evolution of new planets)

The reality is that comets are a complex plasma (electrical) discharge of what I have termed the "solar capacitor". The basics of this concept are detailed in the papers attached in the Appendices, in talks and lectures available on CDs and videos and in other publications. I have explained the basics of this rather complex topic so that even school children understand, so it is my goal here to explain this concept so even the least scientific minded of persons will visualize what I am talking about. It is explained first so that when I explain the effects of comets (especially BIG comets) on Earth and the other planets you will be able to "see" exactly what is happening.

Additionally you will be able to fend off the attacks of the establishment scientists who are sure to attack this book, hoping you will believe them because they are armed with Ph.D.'s and NASA badges. When it becomes clear that anyone can understand these concepts, you will certainly start to wonder why it is so difficult for trained Ph.D. scientists to understand. It will become clear that there are ulterior reasons why they refuse to look at this comet model that was proposed in the 1970's and give it the credit it deserves.

Essentially all the reader needs to know for a basic grasp of the material in this book is that our solar system acts like a large electrical circuit based on the following simple facts (all of which have been confirmed by direct measurements of space craft). 1) Our Sun forms an electrical capacitor (a separation of electric charge as is done by a simple DC battery in a flashlight). 2) This "solar capacitor" has its negative pole at the surface of the Sun, and also has a positive pole far out beyond the outer planets in the form of a sparse "nebular cloud of dust and gasses". An excess current of protons (positively charged subatomic particles) continually generates and supports the solar capacitor by way of the "solar wind" (literally a wind of such particles leaving the Sun and blowing outwards into space).

All stars and galactic nuclei, and even unlit small stars such as our "planets" Jupiter and Saturn are producing these types of cosmic batteries around themselves. This is a result of a natural by product of the nuclear fusion process (the burning of nuclear fuel such as hydrogen, helium, etc.) in the **atmospheres** of these celestial objects. Additionally, there may be many zones and regions in the space that surrounds these objects that complicate these batteries such as rings of dust and gasses, or zones of atomic or chemical materials that have settled into certain areas due to a differential separation in the proton winds.

Astronomers have taken numerous pictures of such clouds that surround other nearby stars. An example is the "Eskimo Nebular cloud" structures around a central star as shown in Figure 1.

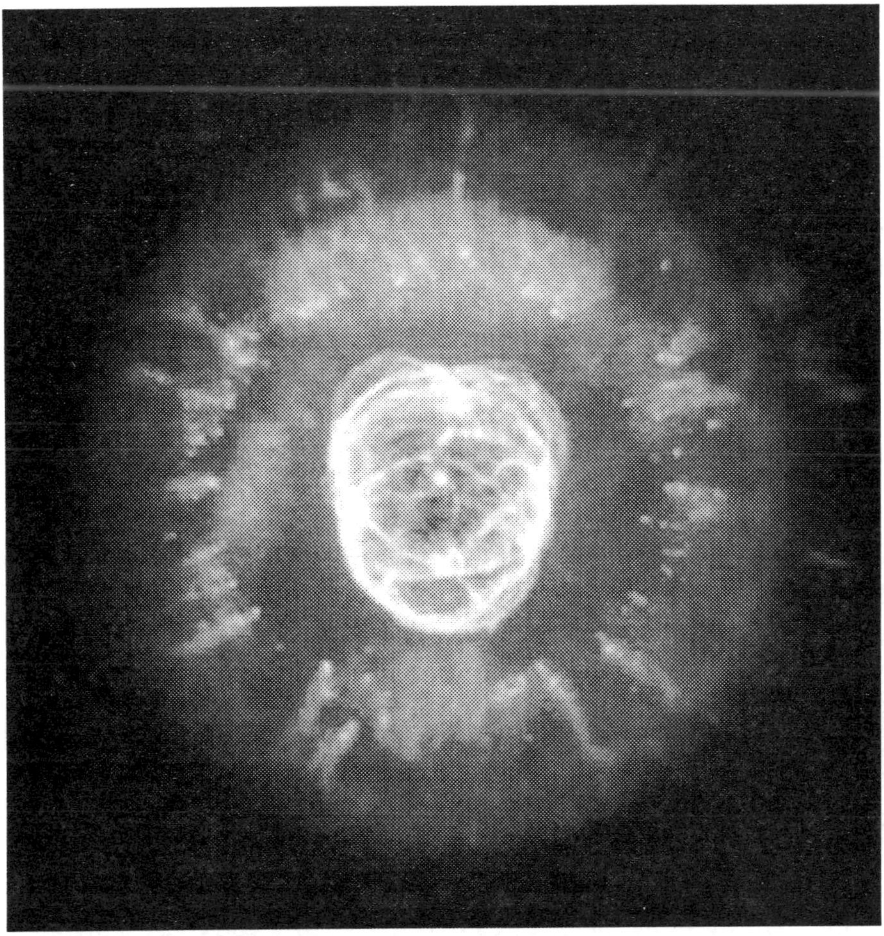

(NASA)

Figure 1. The "Eskimo Nebula" surrounding a nearby star. (The star is the small white dot in the center)

I will give other every day examples of this to help illustrate how this "capacitor" reacts when a rocky asteroid or foreign object comes into this area from the space beyond this controlled region that surrounds our star which we call the Sun. These foreign visitors can become "captured" (a well known process) by the already existing planets and become permanent members of our solar system. This is how the solar system works.

Contrary to the traditional belief that that the solar system formed 4.5 billion years ago and no new planets or moons have formed since, the new reality is that new members are constantly being added and may arrive at any time. Most of these new arrivals will be small but occasionally some may be large or even very large. Some may be in extended orbits moving far out from the inner solar system area and returning every few thousand years. It is this class of object that we see in the large comet Hale-Bopp (to be discussed later in this text) or the object described by the ancient Egyptians they called Nibiru, which they claimed arrived every 3600 years to ravage the inner solar system and our planet Earth.

The further reality is that when these objects arrive, they will enter the "solar capacitor" region around our Sun and ignite a large plasma discharge that will be visible to us as a "comet". The interesting and most important aspect of this phenomena is that the comet is not melting away as in the "Dirty Snowball Comet Model", but is drawing in the vast tail material and is building up and becoming larger. As noted already, most comets are small, but occasionally there will be large or very large comets. These are the ones (such as the anticipated Planet-X) which are growing much larger and pose a serious threat to the existing planets including Earth.

The small comets generally discharge only the small localized regions of the solar capacitor, whereas the very large ones can discharge up to and including the entire solar capacitor, draining much of it's stored energy in a very short period of time. These are the dangerous comets that will disrupt the Sun and cause violent solar storms as they interact with the sun, which will in turn affect our weather on Earth. If they come close enough to Earth or the Sun (even though they may not collide directly with us), they can cause a myriad of violent and destructive events that will be described later in this text.

It is the "drag" of the infalling comet tail material that slows the object down, changes its orbit in an erratic way, and additionally causes the comet's orbit to become more circular with time. Eventually it nestles into a nice neat orbit with the rest of the existing planets, or becomes captured by a larger planet as a new moon. Before it obtains its final state in the solar system there is a good chance that it will encounter the other planets in a wild and reckless game of celestial billiards, in some cases causing major damage to the existing planets.

It is essential first that the reader get a better feeling for the processes that are involved in the evolution of comets (especially the large ones) so that when the resulting damaging "Earth Changes" are described, you will understand them rather than just accept them on blind faith. You will also be able to resist the onslaught of negative publicity that this book is sure to invoke from the "respectable scientific community" whose job it is to protect you from the "truth".

The following is a "talking version" of results of The Plasma Discharge Comet Model, the details of which are found in the Appendices and other sources noted. It should be noted that traditional astronomers continually attempt to mix their incorrect theories with this model and arrive at incorrect conclusions. If you have been taught or are familiar with the concepts of traditional astronomy, geology and weather, then you must step back and begin anew because essentially everything you thought was true is wrong.

The solar system is electrically dynamic. Just this fact alone proves that the Sun cannot derive its heat and energy from deep in its core as suggested by traditional astrophysics theory. This is the topic of great depth and detail, but can be understood from just looking at the surface of the Sun. The Sun is powered at its atmospheric surface by an electrical fire of hydrogen and helium that we call "fusion" which is constantly ignited by energetic lightning bolts in its turbulent atmosphere. It is the local electric field at the outer surface of the Sun (the solar corona of high-energy electrons) that hurls the vast solar flares out into the far reaches of the solar system. The positively charged protons are accelerated outwards while the negatively charged electrons are retarded, thus causing what I have called the excess current of protons in the solar wind. *The Sun produces far more energy in the form of electrical energy than it does in the form of light energy.*

Normally the Earth is buffered and protected from this energy by our extensive magnetic field, but when conditions change rapidly, the Earth is subjected to electric fields and electrical currents that affect our weather, alter our magnetic field, can trigger increased earth quake and volcanic activity and excite cyclonic storms such as hurricanes and tornadoes. Traditional space scientists and astronomers have only spoken in terms of "magnetic fields" in outer space that seem to exist without a source. This is because they are deathly afraid of the terrible word "electric field" which they also claim cannot exist freely in outer space.

Electrical theory is quit clear however, magnetic fields do not self generate and exist without electrical currents flowing … and electrical currents do not flow indefinitely without a battery of some sort to sustain them. Herein lies the downfall of traditional modern astronomical theory. It was the realization of the ability of the Sun's nuclear fusion **at the top of the solar atmosphere** (and not deep in the sun's core) that explains where this battery comes from and how it is maintained. Of course the entire end to end details are far more complex, but this is all that is needed to move on to the dynamics of comets as they evolve into new members of our solar system.

With this in mind, here is how a Plasma Discharge Comet works. An asteroid or rock (large or small) that is zipping through outer space by chance may enter the region of our solar capacitor which we now know extends as a giant nebular sphere or doughnut shape far beyond even the orbit of the planet Pluto. As it enters the solar capacitor it will cause a localized and then more extended electrical discharge of the capacitor. This is much like the back yard bug killer or "zapper" that discharges when a bug flies between the high voltage screens. The physics of this discharge is as follows. Please bear with this brief technical explanation as it will help when we talk about electrical discharges between comets and Earth, between comets and the Sun, and the enhancement of solar and Earth weather.

The electrons (which possess a single negative unit of electrical charge) are approximately 2000 times less massive than the protons (which possess a single positive unit of electrical charge). Thus when a current begins to flow in outer space (as opposed to say a wire in an electrical circuit) both the electrons and protons are free to move since there is no resistance in outer space. The electrons move outwards from the negatively charged Sun and the protons (along with ions and charged dust particles) move sunward towards the negatively charged comet nucleus. This electrical path is always along a line connecting the Sun and the asteroid or object (that is why the visible portion of the comet … the tail … always seems to point away from the Sun). The key to understanding is that the electrons will move 2000 times faster than the protons due to their lower mass. So the electrons will make up the majority of current in this electrical discharge.

In a real comet, this electron current is invisible except when it reaches the fully developed comet coma or cloud of dust and gasses, and may in more energetic comets illuminate the comet and cause x-rays, for example, to the sunward side of the comet nucleus. At the core of the comet nucleus of course will be found the original rock or asteroid that entered the solar capacitor. This is why the small comet nuclei for which NASA has released photos appear as hot dry rocks … because that is exactly what they are. This is also why there cannot be and never will be any ice, snow, water or any volatiles for that matter anywhere on or in the small comet nucleus. Large comet nuclei will gather and hold such volatiles because they have large gravitational fields that will hold these chemicals from floating free back into outer space once they have been collected by the electrical collection process as described in the Plasma Discharge Comet Model.

The ancient Hopi legends speak specifically of what they called the "devil winds" that were caused by comets, and that "the blue comets are the dangerous ones". The blue ones of course we now know refers to the comets with extensive electron currents passing through them causing them to essentially light up like a light bulb. These in turn either interact directly with Earth's upper atmosphere OR interact with the Sun's atmosphere causing solar storms that will in turn send large solar flares our way causing extensive Earth weather and related effects (wind storms, lightning, cyclonic storms, auroras, etc).

The majority charge is carried by the electrons and these accumulate around the comet nucleus, causing the comet nucleus to become negatively charged and therefore it begins to draw in vast amounts of positively charged dust and gasses that are stored in clouds and rings throughout the solar system. Since the electrons move faster, the comet nucleus will always remain negatively charged as it sucks in the material in the ion comet tail.

This is why I call the comet a giant cosmic vacuum cleaner. The comet tail is being drawn in by powerful electrical forces and is being powered by the energy that has been stored up over time in the solar capacitor. Most comets are small, but statistically, a few occasionally can be large or even very large. These are the destructive comets that are on a path to becoming new planets in our solar system.

The most dangerous ones are those in Earth crossing orbits where a close approach would cause the new arrival to interact electrically and gravitationally with Earth. It will be shown that a direct collision is very unlikely and not at all necessary for severe levels of destruction to occur. The Plasma Discharge Comet Model explains why "action at a distance" is the key to understanding the potential damage that can be caused by a large new comet working within the solar system.

> This is why it is essential that as a civilization we must recognize this and cast aside the petty arrogance of the NASA scientists, and face the fact that sooner or later we must prepare, face **AND SURVIVE** the onslaught of one or more of these new arrivals. If we do not stop putting our resources into petty Earth wars and equipment for warfare, as a civilization and species, we are doomed. As this book goes to print there are no fewer than 54 wars being fought on this planet. Providing arms for these wars is the biggest business on Earth. The USA is being moved by a group that has the largest military budget in the history of the world. The President's statement is that he is embarking on a holy war without end and with no cap on the cost or commitment of resources. Nature has only one plan for such a crazed and misdirected ideology.

As the comet develops and discharges the solar capacitor, the increased electrical energy imparted to the solar atmosphere ignites a higher level of nuclear fusion causing the Sun to become excited above its normal levels and may ignite small to very large solar flares. Small comets have many times been observed to directly cause solar flares as they pass near the sun. Yet in light of this obvious data and in spite of my repeated statements that this is what my theory predicts, NASA scientists continue to state that these are mere coincidences and continue to collect their lofty pay checks, receive their grants, publish their papers in the "respectable journals" which are refereed by their friends, and last but not least, incorrectly design expensive space craft to explore these amazing energetic objects.

With this as an introduction, the concept of varying sizes of comets is essential to understanding which comets are the more dangerous ones. Potentially any comet or object moving near Earth or another planet could be dangerous, but small comets like Halleys Comet with a nucleus of only a few miles across produces a relatively small low energy tail. Unless that small rocky nucleus hit Earth directly there is little it would do by passing nearby. In fact Earth passed through the tail of Halleys Comet in 1910 and with no ill affects.

The mistake modern astronomers make is that they assume that all comets will be small and docile and this is simply not the case. Now that we know that the central core nucleus of a comet is not a little snow ball, but in fact is any form of asteroid or rock up to and including giant planet sized objects, you will see that there is always the possibility that a few comet nuclei will be very large. There appears to be a direct correlation between the initial size of the nucleus and the end size of the visible part of the comet which we see as the vast tail that is being dragged ***into*** the comet nucleus.

Additionally there is a difference in what happens to the material that is drawn into the nucleus. The small comets do not have enough mass (and therefore have too small a gravitational field) to hold the volatile light elements that are pulled in and so these disperse as "neutrals" or neutral ions once the ions recombine with electrons in what I have termed the "recombination zone" that surrounds the comet nucleus. Thus small comet nuclei will be found to only have a layering of fine dust on their surfaces, or may be tarry and black due to the electron beam burning the hydrocarbons onto their surfaces as was seen with comet Borrelly by NASA's Deep Space 1 space craft.

Large comets on the other hand that have substantial gravitational fields will be able to hold the gasses that accumulate in the "recombination zone" and will retain these as an atmosphere. There is also a serious amount of heating that occurs in the large comets that is sufficient to cause the developing object to become molten. When this occurs the object will self-gravitate into a round sphere. If electrical discharges occur as it passes other planets or the Sun, then this will be an additional source of heat.

Water is one of the primary components of the comet tail. Water forms in the comet tail from the combination of free hydrogen and free oxygen that are stored in gas rich areas of the solar system. When these two elements combine they give off a tremendous amount of energy in the form of heat. In general, the very large comet is already forming the myriad of chemicals that we find on Earth including oil. Other chemicals are also forming and there is a quite long list of chemicals that have been directly observed in the comet tails. The planet Venus is a classic example of a new planet recently formed from a comet. There is no water on Venus since it is so hot that the oxygen and hydrogen are held in other more volatile gasses.

The evolution of celestial objects from comets is quite complicated and we can see just by observing the planets and moons of our solar system that no two objects are alike. Take for example the topic of moons and magnetic fields of the planets. It has long been recognized that only the planets with moons have substantial magnetic fields. It now can be understood from the basis of electrodynamics of the solar capacitor. All the planets and moons are involved to a degree in the discharging of the solar capacitor. The moons charge electrically just as comet nuclei do. Electrical currents connecting to the associated planet are the factors that create and maintain at least part of the non-permanent planetary magnetic fields. The exact way this occurs of course is quite complicated but the basic result is understandable. In speaking of Earth weather, the phases of the Moon are important, with the passing of the new Moon having the most effect on our weather.

What is interesting here is that unless a planet has a substantial magnetic field, as does Earth, it would be difficult to support life since the planet would be continually buffeted by high energy particles from the Sun. Long term stable biological evolution would be more difficult to achieve (unless all the creatures developed lead shielding on their backs).

Magnetic fields, those invisible barriers that divert the intense and constant solar wind away from our tiny planet are extremely complicated and generally misunderstood even amongst well-trained space scientists. This is because there are many components to the evolution and structures of planetary magnetic fields. Unlike a simple bar magnet that you played with as a child, the planets' interiors are segmented. The iron and nickel metal deposits are distributed in regions and pockets that are scattered throughout the interior. These would never have formed "permanent magnetic fields" if it were not for the external magnetic fields maintained by the solar wind and moons as the metals solidified and cooled over the eons through what is known as the "curie temperature" (the temperature of the metal at which it cools enough to trap and sustain a permanent magnetic field).

Therefore on Earth I have devised a model of the Earth's magnetic field which has at least 5 or more layers or components. The innermost is rather small and is deep within the central core of the

Earth. This of course is where the initial comet nucleus resides that was the initial "seed" of the comet that formed planet Earth a very long time ago (most probably billions of years ago). In the outer crust and mantle of the Earth are deposits of iron and nickel which hold pockets or regions of magnetic fields. Examples of such regions are the Mesabi Iron Range area of Northern Minnesota. For whatever reason, there is a very rich deposit of Iron that man has used to provide much of the world's steel for quite some time. That entire area has strong localized magnetic fields. Compasses are worthless for finding directions in these areas.

So the Earth maintains a general surface magnetic field and one can obtain maps of the "direction of magnetic fields" that cover the earth's surface. In general these "permanent" components of Earth's magnetic field point to the direction we call "north", however the north magnetic pole is far from the true or rotational north pole (the point at the top of the planet around which we spin on a daily basis). Man has located all sorts of anomalies in the permanent magnetic field of Earth and we even have detailed maps of the magnetic pockets on the planet Mars due to satellite mapping of that planet.

There are many other "variable layers" of the Earth's magnetic field. These are caused by a complex interaction of the solar wind with our existing small inner or permanent magnetic field and here is how it works. It is based on the principle that magnetic fields tend to align with each other. As the solar wind blows past Earth, the solar wind particles interact with the small permanent magnetic field. The electrons are forced to bend outwards and around Earth in one direction and the protons move in the other direction. This flow causes the Earth's ionosphere to set up 3 distinct current flows that make up the major jet streams of our upper atmosphere. At the temperate latitudes the positive ions flow easterly, while at the equator there is an electron jet stream that flows westerly or towards the west.

These currents make up what I call the variable magnetic field and as long as the solar wind is constant and smooth, our overall magnetic field is stable and constant in the northerly direction and aligned with the permanent core magnetic field. There is an additional minor component of electrical currents just like that in the ionosphere that is induced in the mantle, crust and viscous molten layers in the interior of the Earth.

Not to complicate this further but there are two external layers farther out in space from the Earth's surface. The first is pair of radiation belts known as the Van Allen belts (rings of high energy particles that circle our globe) and outside that much farther is the outer current of the solar wind itself that adds to the outer shell of the multi-layered magnetic field.

There is one last major factor … as the solar wind rushes past Earth, there is a substantial electron current that drags far out to the night time side of Earth causing what some scientists call the "magnetic tail". Actually when you truly understand that this is an ongoing small comet-like tail of the Earth (with electrons flowing outwards and ions flowing inward from the night side) you must add this component to the complex magnetic field already described.

The major point of this lengthy discussion is that it is important to understand the magnetic field and its association with our upper ionosphere and atmospheric jet streams to understand how Earth weather is affected by solar storms (solar flares that interact with our magnetic field) and cause severe weather at the planet's surface. It is interesting to note that this theory also readily explains the source of electrons that cause the Aurora Borealis or "Northern Lights" (something traditional astronomers are still puzzling over today).

It is an extension of this principle to understand how comets that interact with the Sun and are involved in the electrical discharge of the solar capacitor can also affect weather on our planetary surface. Even small comets have had drastic effects on our weather as we will see later in this book when they can align electrically with Earth, our Moon and other planets.

By now the reader should be gaining an understanding that the NASA theories of dirty snowballs and electrically neutral solar system are radically incorrect and additionally that they will never be able to predict weather or Earth Changes due to comets and other solar system conditions. It is imperative that the public turn to the truth of what is really going on and force NASA to release any data regarding new incoming comets or other large objects. Do not hold your breath however. If the past is any prediction of the future, they will not give up their power positions to the very end.

Understanding the nature of the Earth's magnetic field and the currents that flow throughout the solar system is essential to the correct interpretation of what we are presently seeing. The Sun is currently at levels never experienced before and it is increasing to record levels of activity every day. We were supposed to have peaked in the year 2000 with the "solar maximum" ... we are currently over two years beyond that and there is no end to the increases in sight. It definitely indicates that the Sun is currently interacting with a large intrusion into the solar capacitor.

The topics of Planet X, Comets and Earth Changes will be replete with readers requirement to understand our magnetic field and how it interacts with the solar wind and electrical environment of the surrounding outer space. At this point, after reading this book, you understand more about this topic than all of the NASA scientists combined. The Earth's magnetic field will come into play time and again as you read the chapters of this book.

There was a campfire area found at a primitive aboriginal camp in Australia where the magnetic field under the fire pit was the opposite of that of the surrounding area. This means that during the time of this campfire and as the ground and metals under that fire cooled and solidified (the metals cooled through the Curie temperature), the Earth maintained an external magnetic field in the opposite direction from north or normal. Some have interpreted this to imply that the Earth has reversed its magnetic field and will do so on a regular basis. What this says to me is that the variable component of the magnetic field was temporarily reversed for a day or two as this area cooled, possibly due to a massive solar storm or comet passing near the Earth that occurred at the time the ground was cooling after the fire was extinguished.

This leads into possibly one of the most important concepts presented in this book. That is the concept of "actions at a distance". That is, a comet involved in the discharge of the solar capacitor need not collide with Earth or another planet to cause considerable damage. The fact is that a direct collision would be extremely rare although we are certainly not immune to that given the craters covering the entire surface of our Moon. The more probable cause of damage would be due to effects that may have been started (e.g. a large solar flare) by a comet that might be on the far side of the Sun from us.

In this scenario the comet would discharge to the solar surface causing a significant solar flare that could blow a huge wall of high energy protons our way, causing an alteration of our magnetic field structure and cause electrical and cyclonic storms at the surface of the

Earth, heat up the oceans rapidly (joule heating that today's scientists mistakenly call "el Nino"). The flow of electrical currents from afar could jolt the Earth's crust to induce earthquakes or volcanic activity. This concept is called "action at a distance" and will be discussed in more detail in the chapters dealing with Earth Changes. Many people have begun to make these relationships and associations, however this book will explain exactly how this occurs and how the processes work and what the energy levels could be in mild, moderate or extreme cases.

Another aspect to "action at a distance" is what I call the "physical" extension of outside factors. There are numerous types of physical effects such as gravitational waves caused by a large object passing near an existing planet that could set up a "tidal wave" in the mantle of the planet and cause severe earthquakes, mountain building and pole shifts. This would be in addition to the ocean tidal waves that would inundate the coastal areas of continents. How devastating this would be is controlled by a number of factors including the size and density of the passing object, the velocity of passage and the proximity of its closest approach. Some people are claiming to know where certain effects will or will not take place, but as this book will explain, apart from ocean tidal waves, it is virtually impossible to predict the location of most effects.

Another "physical" effect would be exactly what we observe has happened with the planet Mars. Mars is observed to have had oceans in the past and these are gone and the atmosphere that held the water and oceans and rivers in place is also gone. Why would a planet nearly the size of Earth with vast oceans and a substantial atmosphere holding it in place all of a sudden have it removed? NASA scientists without any explanation place this in the distant past (mainly because they have no clue as to why it happened so that way they do not have the explain anything to the public). The reality however is that if this had happened millions of years ago, the vast dust storms that ravage Mars every year would have erased or at least eroded the geological features (e.g. delicate meandering river beds) long ago.

Additionally Mars shows no cratering where the oceans were, yet is littered with craters where the land areas were prior to its being relieved of the oceans. This means that as the craters were made, the ones under the oceans were eroded away whereas the ones on land weathered more slowly and still remain. In reality, Mars could not have lost its atmospheres more than a few thousand years ago, and there are historical records that the ancients saw a large comet pass by Mars and lift off the oceans and atmosphere in a single night. This aspect of "action at a distance" works as follows.

If a large comet nucleus comes near enough to a planet to interact electrically, that is, the comet establishes an electrical connection with the planet, then what is called the "surface gravity" of the object is allowed to take over. If this happens the object with the larger surface gravity (the force of gravity at the surface of the object) will suck at the volatiles (air, water, etc) and if this process is allowed to work long enough, the object with the larger surface gravity will remove most or all of the atmosphere and oceans of the smaller object. One object is stripped of its cloak and the other receives a significant pollution event. It is apparent that Mars recently underwent this "physical action at a distance" and we see its pristine river beds unaffected by eons of dust storms staring at us in the satellite photos of our neighboring planet. This also proves that the object was at least larger than the planet Mars (e.g. as large as the planet Venus or larger).

Earth has in fact been involved in similar events but so far we have managed to "win" the tug of war … receiving what amounts to vast pollution events (the rain of brimstone and burning naphtha or hydrocarbons as well as water flooding). It is clear from this also that the world's oil did not come from millions of years of decaying fern forests (no chemist has ever developed a process to convert decayed matter into oil).

The oil fell onto the Earth during such a pollution event when the Earth performed celestial battle with a large cometary intruder. The great earth wide flood was also the result of the water in the comet tail falling into Earth during the same or similar encounter (remember water is the primary component of comets). Once again most scientists do not realize that comets can become very large and that the volume of water, oil, etc. that is produced in the tail of a very large comet would dwarf the atmosphere of planet Earth. When the oil in the Mid-East was first used in quantity it was laying in pools on the ground. When this was used up, they dug into the sand and now we are drilling into oceans of oil that seeped into the sand after falling from above during the many comet encounters that Earth has seen.

It has always amazed me when you look at the tar pits with dead animals and then you hear the standard explanations that thousands of animals supposedly just walked into this tar pit and one by one died!!!!!!!!!!!!! What really happened was ... all over the world the hydrocarbons rained into Earth and formed oil lakes and pools over the Earth's surface, killing the animals and drowning them in lakes and rivers of oil. As the oil was absorbed into the Earth, the tar pits remained where the animals were preserved until today. Even human remains have been found in the tar pits.

There are other forms of action at a distance that occur and these will be discussed at length later in this book.

Another aspect of the arrival of new objects from outside of the solar system is that they many times arrive in multiples. This is because whatever the source of these wandering objects from beyond, they were formed or fragmented at the same time and have found their way to our tiny corner of the Universe. As soon as they enter the solar capacitor they ignite into a comet and begin to slow down due to the tail drag (this is why they appear to scientists to be objects on orbits that originated near the solar system). They may be orbiting each other OR they may have long since split apart so that one will trail the other by many years or decades. This is common even with comets that are known to exist in the inner solar system. That is how the enormous comet Hale-Bopp entered the solar system (with an orbiting companion). By the way, the nucleus of Hale-Bopp was at least as large as our Moon and contrary to NASA's initial disinformation campaign, Hale-Bopp did have a large orbiting companion (visible from small earth-based telescopes). Only NASA with its hidden data knows where it went. With a simple equation and given just a few time lapse frames from NASA's hidden data showing the orbiting object, one could calculate the mass of the nucleus of Hale-Bopp.

Although there are many subtle and diverse aspects of comets that will not be covered here for simplicity sake, one final topic remains to be discussed relative to the Plasma Discharge Comet Model. It is a subtle aspect I have termed the recombination zone. This is the area where electrons neutralize incoming ions and dust particles and is a rich blending of chemicals and raw materials from which the new member of the solar system is formed. This is where hydrogen and oxygen form water and this is where hydrogen and carbon atoms form hydrocarbons such as oil or what the ancients called naphtha. The smaller comet nuclei will retain only the dust and heavier materials, whereas the more massive comet nuclei with significant gravitational field's will additionally retain the chemicals, gases and volatiles that are synthesized in this cosmic blender. Since comets are not dirty snowballs, the much-publicized concept that they would carry life-bearing molecules to Earth cannot be true either. What would be true, is that the fledgling planet that has developed from a comet, would contain all the ingredients necessary to begin the evolutionary process on its own. The introduction of live from afar theory begs the question that if life is arriving from afar, how did it evolve there? ... it had to start somewhere and by some means, whether evolution or otherwise.

There have been many authors that have written about the subject of planetary catastrophes, but one person stands head and shoulders above the rest. His name is Immanuel Velikovsky and at the end of the main text, I have placed a special tribute to this man who withstood the ravages and ridicule of the scientific communities, but who is now proven to have been very correct about his interpretations of celestial events as recorded by the ancients. This book shows how traditional astronomers failed to look beyond traditional theories, and that there is an entirely different theoretical structure that is correct and shows how Velikovsky's claims were certainly plausible.

Even the most conservative of astronomers is now embracing the fact that the Earth has been ravaged from outer space on multiple occasions. These scientists maintain however that we would have to be hit directly by a comet nucleus or asteroid to receive any damage. Hopefully you are becoming aware that "action at a distance" is the key to understanding the true nature of celestial catastrophes. The advancement of The Plasma Discharge Comet Model replacing the Dirty Snowball Comet Model (with its entourage of incorrect associated theories) must be seen as the downfall of modern astronomical theory and the true understanding of the history and development of the world.

II. ABOUT THE COVER FIGURE

The cover figure of this book is a photo of an Indian rock painting from Green River, UT. taken by photographer Gary Smith. The picture was originally used in the 1980's publication of my 3-part comet paper that is reproduced in Appendix 2 of this book. It is important for the current text as it shows that the primitive artist distinctly drew two comets with squiggly sunward "spikes". The animal-like creature represented at the left appears to be in the state of surprise, holding his arms out as if to show enormous size. The priest-like robed character in the center with the big eyes appears also with his arms extended and with a snake like creature leaving one hand and moving towards a circular object in the skyward direction. Could that represent a lightning bolt being hurtled across the heavens from Jupiter to the planet Mars?

We could speculate about this, however the two comets with the helical sunward spikes definitely represent large comets with spikes in the sunward direction. Using the Plasma Discharge Comet Model, we now understand that these comet discharges were so energetic that they would follow a helical type of pattern as the electron beam (the invisible part of the comet current sheet) gyrates in a large magnetic field that surrounds large comets. Actually Hale-Bopp, also had at times as many as 16 helical arms of triply ionized oxygen and doubly ionized hydrogen (making water) ... again something the Dirty Snowball Comet Model cannot explain. This is due to the large electrical currents moving in a high magnetic field that surrounds the huge comets such as Hale-Bopp.

What is most interesting about the two comets in the cover photo is that the long stringy comet on the left-hand side shows no central nuclear type object, whereas the one on the right distinctly shows a large circular central object. This is just as it would be with a very large object as the central core object of an incoming comet. This was the case with the very large Comet Hale-Bopp when it first arrived in the solar system and ignited into a very large comet. With an extremely large cometary nucleus, say for example, one that might be the size of the planet Venus or larger, the nucleus on occasion would be visible through the cometary coma and sometimes it would not. It is as if this primitive artist meant specifically to leave us this record so that today we would be able to recognize this message for what it truly signified to the artist.

The region of Green River Utah where these drawings are found is unique in that it is a stark almost uninhabitable arid region of the American southwest. In the remote hidden cliff area where these paintings are found, it appears that this was quite possibly the type of sheltered area that housed an ancient group of people who survived the onslaught of a ravaging comet. The adobe apartment-like structures that the residents lived in were tucked up under the cliffs in chasms in the Green River area. The residents that lived in this cave-like setting existed on meager farming and possibly domesticated animals for food.

One aspect of the Plasma Discharge Comet Model that was not stressed earlier and one that separates it again from the Dirty Snowball Comet Model, is the characteristic "sunward spike". First observed by modern astronomers with Comet Arend-Roland in 1957, the astronomers as usual found an explanation that seemed to fit their dirty snowball theory. According to their explanation, this only "appeared" to be a sunward spike and was "actually an optical illusion of a anti-sunward Type II tail". To the layman this means that what they were saying was that they hypothesized this Type II tail (a thin secondary tail sometimes observed in comets that points away from the Sun) to be slightly visible to the side of the main wide (Type I) tail and thus really was not a sunward spike. Comet Arend-Roland and the issue of spiked comets is discussed at length in the papers found in the Appendices of this book.

When I first proposed the new comet model in the late 1970's, as a well-versed scientist, I created a list of experiments that would definitely distinguish the two comet models and prove or disprove one or the other. I made one specific test of using the sunward spike as a critical observation that had to be made. I proposed that sooner or later we would see a comet at a right angle (e.g. from the side and not from a front angle as was the case with comet Arend-Roland).

As luck would have it there was a bright comet (comet Kohoutek) that passed near the Sun and was at a perfect right angle (broadside) configuration. It was extensively photographed by Space-Lab astronomers in 1974 and the photos showed the full development of that comet's sunward spike. In that configuration there was no doubt that the spike was extending in the sunward direction. As has occurred time and time again however, NASA scientists ignored the data and never bothered to try to give a dirty snow ball explanation. Since then dozens of comets have been observed to have definite sunward spikes, and as we have seen, some have even connected electrically to the Sun causing major solar eruptions. Repeatedly NASA claims that there is no connection between the comets and the Sun's activity. I call this "sweeping the data under the rug", something NASA has had to do all to often to protect its outdated theories.

An additional aspect to the Indian art rock painting is that the ***helical*** spike adds a level of complication. The fact that it was drawn twice confirms that the artist fully intended it to be that way. The only way such a spike can exist to the sunward side of the comet nucleus is for the electron beam, that is an essential part of the Plasma Discharge Comet Model, to be so intense in the comets depicted that it "pinches" into a fine narrow beam … and secondly that the local magnetic field is so intense that it causes the electron beam, that is igniting the comet up like a light bulb, to flow in a helical pattern.

The term "pinch" refers to the property of a free flowing current to pull itself together by self induced magnetic fields to produce a narrow beam such as is observed here. The problem for the Dirty Snowball Comet Model is that it has no strong currents and suggests only weak magnetic fields. It especially has neither of these to the sunward side of the nucleus. Of more importance however is that the currents must be in the Mega-Amp range (millions of amps of current). That is why NASA space craft have only found hot dry and even charred rocky surfaces at the nucleus of the small comets they have visited. That is because that is exactly what they are! In fact it should now be clear why there could never be any ice or snow on a small comet nucleus.

This primitive artist's drawing was chosen for the cover of this book for many reasons. It represents everything this book stands for as it sends a message that encompasses the many aspects of this book. It shows the development of a truly HUGE comet, one that we have not seen in hundreds if not thousands of years in the inner solar system. Its shows the helical sunward spikes that can only be explained by the Plasma Discharge Comet Model. It shows that sometimes the comet nucleus can be visible and the fact that in the comet on the right it shows a distinctive ROUND nucleus is of great significance also. Small comet nuclei are never visible within the dense comet comas (in fact for ages astronomers even questioned whether there even was a comet nucleus because they could never see them).

The point of the round nucleus is that the only objects in outer space that become round are very large planet sized objects which have heated to a molten state and additionally have sufficient mass (therefore have high gravitational fields) that will pull the molten body into a sphere. So the fact that the artist drew the nucleus at all is indicative that he could see it in spite of the fact that it is obviously a huge comet with a fully developed tail.

The picture represents the main premise of this book ... that the ancients witnessed planet(s) forming from huge comets, and have attempted to tell future generations of the damage they cause. In the book "Storms from the Sun", recently produced by NASA to counter this author's efforts, they made fun of those who suggested that comets could be anything but docile little harmless snowballs. It is now at the point where no one believes them anymore.

Another aspect of the photo is that it was painted by a primitive artist who was skilled in the art of drawing. The people in this obvious survival area, from what I can estimate, were there to protect their families from something. That something was the huge comet(s) that were ravaging the solar system at the time. This area would have been an ideal area to protect the people from the onslaught of meteors, burning naptha, flooding, high energy proton streams and other disastrous effects of the presence of a huge comet working in the inner solar system. There is a clear and emerging awareness today amongst the world's populous that there is again an impending object at the far reaches of our solar system and that it may arrive in the upcoming decade. Here is a living example of a primitive population in survival mode from forces that were beyond their control.

There are two comets shown in the drawing. Either there were actually two comets OR this is the same comet depicted twice. It has always been my contention that there were two comets in the solar system at the time of the Great Flood (the last world-wide conflagration wrought by a great comet of destruction). There are many reasons for believing this. One is that the great comet Hale-Bopp was here at just that time and its orbit also suggests that it could have been affected by the planet Jupiter to separate from the other piece that was the one that actually affected Earth.. The reasoning here is too complex and unrelated the science aspect of this book but I did want to mention this to extend the idea that when BIG comets come into the solar system, theory suggests that many times they will come in multiples or have roaming companions that may split off and become independent as they follow their paths and encounter the many planet sized objects already orbiting the Sun in our solar system.

The final reason I chose the Indian rock painting for the cover is that I wanted a striking contrast image that would reproduce well in half tones print. This sets the image I wanted to present of the ancient world meeting the modern world. It is a figure that I never quite get tired of looking at. It causes me to reflect over the centuries and millennia to a time when scared humans huddled under the cliffs of Green River Utah and found hope to survive and go on to raise their families in spite of odds far beyond anything that we could imagine.

III. HISTORY OF THE SEARCH FOR NEW PLANETS

In this book there will be no reference to a specific Planet-X nor will it make any predictions of future events or make concrete statements about dates of past events. Although I have my own views on all of this, this book is meant to be a guide to the scientific explanations of how the electrically active solar system works. It also attempts to put this very complex topic into perspective. For that reason it would be beneficial to put the entire topic of the search for Planet-X into perspective.

The Term "Planet-X" is used in many different ways by different authors. This text uses it in the generic sense. It refers to possible devastating comets of history or possible future new entries into our solar system in an effort to explain the myriad of Earth Changes that might be anticipated in the new understanding of how the solar system works. To others, however, it refers to a particular massive "planet" or other object(s) that would have, for example, a historical repetitive cycle (such as the planet of Egyptian lore called Nibiru), and is claimed to have been the cause of extensive damage to Earth.

The search for "Planet-X" goes in both directions on the time line of the history of the Earth and human species. Certain authors have already looked into the past and found ample evidence that our planet has been "blindsided" in the past, and they are expecting that the future holds the same.

The current book takes the generic attitude that … it is clear that these events have happened in the past, and they will again happen in the future. The question is not "IF", but only "WHEN". ***The real answer to this is very complex***. Even if the exact location and orbit of a new large incoming object were known today, ***it would be impossible to tell exactly when and where it would be in a month or even a few days since the orbits change radically on a daily basis due to the variable comet tail drag.***

The ongoing current search for Planet X is in a great part based on evidence of past events. It is interesting where the centers of interest exist today. Even the Vatican is hot on the trail of searching for Planet X and has paid for and is staffing large professional observatories. ***What is clear and can be said with 100% confidence is that NASA***

has observed such objects and is hiding the data from the public. They are under strict contracts to do so.

My personal role in this process is one of an independent scientist (not subject to pressures of the scientific community, peer pressure or government non-disclosure agreements and funding) who will explain the myriad of possible Earth Changes and their cause. I leave it to others to research the specifics of confirmations of sightings and speculations thereof. This separation of function works well as ***my theoretical work does not depend on a specific date or time or event, but covers all time that such encounters may take place, whether past, present or future.***

The ancients we are told knew only of the planets out to Saturn. The reigning astronomical academic community believes that all nine of the planets were created in a swirling ball of dust and gasses 4.5 billion years ago and that no new planets have (or could) enter and form in the solar system. The entire thesis of this book is that they are wrong and that there is ample evidence historically and in modern data that unquestionably shows that new members are constantly arriving. Although most new arrivals are small (e.g. small asteroids and comet nuclei), some can be very large of the size of the major planets and moons. NASA has projects to look for the so called Kuiper belt objects, yet even the properties of these do not match ancient cold bodies that would have existed for 4.5 billion years at the edge of the solar system.

It is further stated here that the varying ages of the members of the solar system should be self-evident by just looking at these objects of which no two members look alike. The most startling contrasts come with the moons of the giant gas planets Jupiter and Saturn. I find it impossible to look at these and imagine that somehow they formed at the same time. Much of the theoretical work performed by space scientists and astronomers over the past 30 years has been in thinking up excuses for why they solar system has not turned out to be as their theories had predicted.

Let me explain how this "process" works (something I witnessed countless times personally). As the unexpected data arrives via satellite from the far reaches of the solar system, the amazed and bewildered scientists sit around and throw out ideas. In many cases the contradictions are glaring (as in 100% contradictory to traditional theories). As the ideas are bantered around, and a "story" has to be given to the awaiting news media or journal rep, a straw vote is taken and that idea sticks ... many times for good. Eventually the contradictions come out, but by then NASA has its pat answer or apology as I call it, and the rapidly made up "excuse" is re-sung constantly until everyone believes it must be true. Examples of this are the high temperature for Jupiter's moon Io, the electrical currents running from Io to Jupiter, the high temperature of Venus and the fact that the alleged Greenhouse Effect does NOT account for it, the immense size of comet Hale-Bopp's nucleus, x-rays from comets, and the list could go on for pages.

Much of the trouble with these communities is that the vast majority of the scientists are not well-trained physicists. They instead are well versed in the standard explanations for how the solar system functions, and this in fact has created a collective mental block to change or allow new ideas to be considered. The other major infrastructure problem is that these scientists cannot and will not criticize their peers. But there is another side to the coin (in defense of these "professionals").

There have been many predictions made by non-scientists regarding the arrival and effects of a new Planet X, and I am sure that this frustrates professional astronomers as much as it does me. There is so much well-intentioned misinformation floating around that it becomes difficult for the average person to sort it out. I hear people quoting this misinformation all the time and I realize the level of scientific savvy of the general public is limited. That is why my lectures have found audiences at all levels of interest, with everyone from retired housewives to scientists with multiple Ph.D.s. People understand what I am saying and it makes sense. That is why this book is so unpopular in certain circles also, it is exactly what they do not want (a well educated public that can think for themselves on these issues).

Historically, there have been objects seen in the inner solar system that were large and observed by astronomers. An excellent account of these is contained on an internet web page which covers the "search" going back hundreds of years. It also includes an excellent history of the search for Planet X which began with professional astronomers when they noticed that the movements of the outer planets did not follow the predictions of Kepler's Laws of Celestial Mechanics. That page covers the topics of Vulcan, the intra-Mercurial planet,

Mercury's Moon, Neith the moon of Venus, The Earth's second Moon, the moons of Mars, the 14th moon of Jupiter, Two new moons of Saturn, new moons for Uranus, the search for Planet X and Nemesis (the Sun's companion star).

With such an excellent reference at hand it would not serve this book to repeat those details here, so the reader is referred to the internet web site

http://seds.lpl.arizona.edu/billa/tnp/hypo.html

Since the majority of people today have access to the internet or to a public place, the reader is referred to this excellent review.

Regarding the current search for a return of the historical planet Nibiru, it is interesting to note a Vatican memo to NASA which asked the question, "Is Hale-Bopp Wormwood?". This refers to the old testament reference to the terrible planet called Wormwood that poisoned the waters of the Earth. The plant name "wormwood" which is considered an herb by some and a poison by others is related to this biblical reference.

Since comets come in multiples as explained elsewhere in this book, could it be possible that the massive comet Hale-Bopp was a precursor or companion with a larger and more destructive planet called Nibiru. It is my opinion that there were multiple large comets in the solar system when the ancients saw the "Sun stand still". One possible interpretation for this passage is that … through the haze and clouds of pollution that engulfed the Earth during a passage of one of the HUGE comets, the glowing comet nucleus was mistaken for the Sun. Since it was in the same part of the sky as the Sun, it was mistaken for the Sun.

The Search and identification of candidates for Planet X will be the ongoing topic of radio and other broadcasts by this author as it is the ultimate goal to make this identification. At that point the estimation of potential damage can begin. Once again, do not depend on the explanations of NASA since they do not understand the true ramifications of the arrival of such an object with their incorrect and outdated theories. No one within NASA has the proper background to make such judgements since they have repeatedly shown that they do not understand the true nature of the Plasma Discharge Comet Model.

IV. THE AXIOMS OF ASTRONOMY

An "axiom" is a predetermined initial assumption that appears to be reasonable and correct. They are sought after by mathematicians to form the logical basis of complex mathematical systems. One of the most famous of all is the set of axioms called Peano's Postulates. These few and extremely simple postulates that seem to be almost self evident, are used to create the basis of the subject Logic which in turn forms the basis of all of mathematics. To mathematicians it gives them the ultimate sense of security that everything in their mathematics is sound and structured, and most of all, is based on these few and self-evident axioms.

In science there are many "laws" such as the first three "Laws of Thermodynamics" as they are called. There are the "Laws of Conservation of Energy" and "Conservation of Momentum" that basically state that energy cannot be created or destroyed, but can only change forms. They affect all fields of science and are so fundamental that they form a good part of the studies in Chemistry, Physics and even Biology. The astronomers are no different. Many have spent their entire lives and careers researching and publishing with great confidence that they have reached a set of "axioms" which are sound and can stand as the corner stones of how we think and theorize about the universe we live in.

The problem however with the axioms of astronomy is that they were formulated prior to the exploration of the planets and the invention of more sophisticated methods of measuring the space around our solar system.

To put this into perspective a bit of history is necessary. In the late 1940's the world of science was still developing. All of a sudden a series of 3 books hit the international market. They became international best sellers and were translated into 14 languages. They were written by a man named Immanuel Velikovsky and titled Worlds in Collision, Ages in Chaos and Earth in Upheaval. In brief, Velikovsky claimed that the Planet Venus was in historical times a massive comet that ravaged Earth on two successive occasions and it soon nestled into its current orbit. He claimed that Venus was observed by the ancients all over the world to have "come out of the planet Jupiter" and also passed near Mars where it's serpent-like extension sucked the oceans and atmosphere off of that planet. The entire story of these books and the astronomical community response has been accounted in extensive writings by hundreds of authors on both sides of the issue.

My theoretical work and this book show how such a giant comet could in fact be "captured" by Jupiter so as to appear to be "born of Jupiter", and how it's electrical discharge plasma connection ... assisted by a large gravitational field ... could in fact pull the oceans off of Mars and create the Great Flood on Earth and associated catastrophes noted by the ancients as retold by Velikovsky.

Be that as it may, the news media in 1950 went to the astronomical community in the wake of Velikovsky's books and asked what the "real" story was. As it turned out, they really did not have a comprehensive all encompassing theory of how the universe or solar system functioned, although there had been many attempts at giving varying explanations. Especially, comets were poorly understood. A young Harvard professor named Fred Whipple proposed what became known as the Ice-ball Comet Model and eventually the "Dirty Snowball Comet Model". One of the main criteria was that comets would only be small and were melting away (they certainly could not be huge or become planets as suggested by Velikovsky).

The point is that all of these theories that were decided upon at that time still exist today forming the basis of astronomical theory. Velikovsky's books remained best sellers throughout the next two decades. The world was buying them and reading more than any other books. It was the best selling book that Doubleday ever had. This progressed into the 1970's until a young aggressive astronomer named Carl Sagan (who held the David Duncan Astronomy Chair at Cornell University) took up the banner. He organized the many academic communities and led the charge against Velikovsky.

Few people know that Sagan gained his notoriety (and eventually the lucrative host position of the COSMOS television series as well as many book contracts) because of his efforts to publicly denounce Velikovsky. What also is interesting is that Sagan at the height of his popularity, applied to become an astronomer at Harvard and was rejected because he lacked a track record of serious original work in the area of astronomy. At any rate, the astronomical community stood solidly behind the newly found "Axioms of Astronomy", and never once attempted to analyze the obvious … could there be another set of explanations, and could their holy set of axioms be flawed?

With the burial of Velikovsky accomplished, astronomers hoped to put this nightmare behind them and get on with life as "normal". It was at exactly this time that I happened along at Cornell (1979) just after the public denunciation of Velikovsky (1974). At the time I had never heard of Velikovsky as I had been out of the country for quite some time and was unaware of the hornets nest I was about to step into.

The preface of this book outlines some of the history of the early days of the theoretical work and my efforts at presenting it and publishing as a young new faculty member of that lofty university. You might imagine that this entire scientific community was just sighing a breath of relief, when a new young member of their very own community came along and began publishing his theories stating that comets were not dirty snow balls, but were in fact planetary objects in their formative stages.

Not to make this bad enough, he was using the latest data returned from the space probes they designed to prove it. This is an entire story in itself (I hope to write a book on that experience some day). As I watched and observed the influx of Voyager I and II data from Jupiter and Saturn, I recognized my theoretical work in that data. I also witnessed the continual denial that the NASA space scientists went through to try to force square pegs of data into the round holes of their theories, a practice that has become standard practice at NASA up to and including today.

After the 3-part comet paper (reprinted in the Appendices of this book) was published in the early 1980's, I had a statement that I wanted to make. It came in the form of a short article that was published in the same journal KRONOS that published the comet paper for posterity. That short essay I called "The Axioms of Astronomy". It is reprinted here to put the upcoming parts of this book into perspective.

Original appeared in:

KRONOS – A Journal of Interdisciplinary Synthesis X:3

Summer 1985 ISSN 0361-6584

THE AXIOMS OF ASTRONOMY

To the Editor of KRONOS:

A renowned British physicist recently wrote to me concerning his confrontation with editors of a popular astronomy magazine. He stated that "there is far too much evidence of suppression of papers which do not take the now orthodox Oort-Whipple (ice ball comet) model as an axiomatic starting point. . . . It does make one wonder if Astronomy qualifies as a science!".

In 1980, in a letter of rejection to a paper submitted for publication which predicted electric fields near the planet Saturn (the paper was subsequently published elsewhere), a journal editor scolded me for not following "the scientific method". He explained that one must start with accepted theory as presented in the respected journals, reference it properly, and add any new work to this frame-work. He added that the workings of Saturn's ring system were well understood and that there was no need to consider electrical phenomena. Four months later, Voyager I reached the electrified planet Saturn and sent back what some scientists consider the most mishandled data in the history of modem science.

What are the axioms of Astronomy? They are the basic concepts around which theory and data must conform as seen by traditionally educated scientists. They are:

- the solar system formed 4.5 billion years ago and the order of the planets has not changed since (first stated by the mathematician Laplace).

- the universe began in a "big bang" about 10 billion years ago and the universe has been expanding ever since (first stated by Hubble at the turn of the century).

- Gravity is the sole force that governs the cosmos (first stated by Newton).

- Comets are frozen balls of ice which melt when near the Sun (first stated by Kepler).

All other "accepted" theories of Astronomy such as the Venus Greenhouse Effect, and scores of others, as well as theories in Geology (i.e., plate tectonics), Archaeology, and Evolutionary Biology have been developed to conform with these axioms. Literally all data from space probes are interpreted by NASA scientists to support the axioms. The popular science magazines and newspapers force-feed the public with scientific hype, constantly referring to these axioms.

Using the axioms, a great many predictions have been made about the expected state of the cosmos. But what happens when data from new work or research contradicts the expectations of the axioms? There are three possible courses of action: 1) reconsider the basic axiomatic structure, 2) rationalize the data to support the axioms by creating after the fact ad-hoc theories, or 3) sweep the data under the rug. Unfortunately, the peer editing systems of our scientific journals only support the last two possibilities. There is a reason for this.

These axioms have formed the cornerstones of astronomical belief at least since the early 1900s. The fonts of western knowledge lie in the universities where tradition and the respected scientific journals are based. The strict hierarchy of advancement (from graduate student to aged tenured professor) forces underlings to comply with this protocol. Dissenters and original thinkers have no place in this structure. One is only granted permission to build on the pre- existing caste system. Change cannot be expected, however necessary or justified, within this framework.

A thorough look in the astronomy journals shows that they are filled with papers based on the standard axioms, but the axioms are never questioned. How embarrassing might it be if it were discovered that 80 years of published papers were erroneous (especially after censoring and publicly ridiculing those who may later be viewed as correct)? Every grade schooler has heard how the now heralded discoveries of Copernicus, Galileo, Kepler, Newton, and even Einstein (as well as countless others in other fields) were met with closed doors by the seats of knowledge of their day.

Secondly, this is a closed system. These groups reign as "the experts", generally controlling the flow of publishable material to their journals and the popular media. They do not allow those in other fields to cross the interdisciplinary boundaries; they control the new personnel who enter the field and those who seek employment in their field. They are approached by the mass media to inform the public, and form core groups such as NASA and the American Association for the Advancement of Science. They control who is allowed to referee articles to be published; they receive the government grants to perform new research, and they present the results of their research in the journals which are controlled by their peer groups. It is a closed system. There are always exceptions to the rule, but in this case, there are very few.

Anyone outside the system who dares to contradict is easily labeled a non- expert (commonly name calling occurs such as crackpot, pseudo-scientist, etc.). Who is the man on the street to believe?

Insiders who defect can be singled out and eliminated. No one will offer public support since the same fate may befall them.

It should not be surprising that the axioms of Astronomy have remained unaltered over the course of this century, even in spite of remarkable finds of recent space probes. One can only anticipate that future finds from space, including those of upcoming comet fly-bys, will yield the unexpected. . . but watch with precognition as the experts will claim them to support the axioms of Astronomy.

J. M. McCanney (reprinted with permission of KRONOS press)

As a short addendum to this brief bit of history, it should be pointed out that the astronomy / space science community has been sitting on its laurels for far too long and there is a "right-sizing" that is necessary regarding the topic of the Axioms of Astronomy. There are probably many younger astronomers and space scientists today that know little or nothing about the true history of how all of their standard theories came to be, and the controversy that has raged behind the scenes for decades between people like myself and the now elder generation of those hallowed academic communities. This treatise is as much for them as it is for the general public.

It is time to pull the soft wool from over your eyes and as my uncle Ben used to say, "open your big brown eyes and see what's going on around you in the world"!!!!!!!!!!!!! (usually stated with a healthy dose of sarcasm). As many will take offense that they could possibly have been hoodwinked, and their hard earned educations fraught with misinformation, my uncle's words were meant for you. The university walls where you earned your degrees were as much a fortress from getting a real education, as they were to cloister your minds in the correct religions of modern astronomical thought (which is more of a religion than most religions).

The reality of the Ph.D. educational system today, is that if you truly were a true free thinker, you could never possibly get a Ph.D. in astronomy or space science ... since you would know about this history ... and when you brought it up and pushed for the truth, you would have been lead to the door and another eager unsuspecting young mental athlete would have taken your place. Getting a Ph.D. is based on one's ability to pass tests and regurgitate what you have been told. Questioning it is not part of the process. And once you get that Ph.D. and get your first post-doc position and then your first tenure track position, you are well trained in the art of not questioning the axioms of astronomy.

There are also many astronomers, astrophysicists and space scientists who are daily edging towards the realizations that this author held over 20 years ago regarding the electrical nature of the solar system, but was not allowed to state or publish them. Can you imagine where these fields would be today if the last two and a half decades were not spent marching down the wrong road?

The final step in this process of making the theoretical connections in what today is commonly called "The Sun-Earth Connection" is the realization that all the basic theories or "Axioms of Astronomy" are false as noted in this 1985 Letter to the Editor. The first step is to admit that comet nuclei are not out gassing or producing water as was confirmed when the DeepSpace 1 space probe arrived at comet Borrelly, or when the 7 comet pieces of Shoemacher Levy 9 crashed into the far side of Jupiter, with their bright red SO2 tails that were hours before water tails, and which released more energy than all the nuclear bombs in the Earth's war arsenals as they connected electrically with the upper ionosphere of Jupiter.

NO …COMETS ARE NOT DIRTY SNOW BALLS. And when you see what they really are, and realize that the axioms of astronomy then fall as dominos in a row, you will have taken the first step to view where this author stood 25 years ago.

Not until you reach this conclusion and accept it … **AND IMPLEMENT IT** … will the final state of knowledge be achieved in the understanding of true nature of "The Sun-Earth Connection". Specifically, there is no such thing as "magnetic reconnection" or "solar pick up ions" or "magnetic bubbles" … there is no such thing as a magnetic field existing of itself in a plasma, nor is there any possible mechanism for a magnetic field to induce energy into a free particle in a plasma (ala Maxwell's Equations). There are only more fundamental electrical effects, which in turn cause the magnetic effects that are incorrectly thought to be the fundamental effects. The Chapman Physics scenario is fundamentally flawed.

This will anger a lot of professional scientists, and many will at this point throw down this book and close their minds. There is an expression in science when these types of road blocks occur … that the current generation of scientists must retire and die off before the new important ideas will become accepted and noted in their proper places in the history of science.

V. PUTTING NASA INTO PERSPECTIVE

There clearly is no love lost between myself and the people at NASA. But over and above a simple grudge match between us that spans a few decades, I am sure they are somewhat amazed and wondering why I keep "hanging on" and do not fold in my cards as have the rest of the "independent scientists" who have left the system and never returned. They must wonder why I pay my own way to scientific conferences only to be continually placed on the back burner and kept from meeting the news media at planned news releases at the conferences. The simple answer is that I have always known I was right and that eventually this reality would be realized (and when that happened, I fully intended to get credit for it).

They are generally (although not always) polite to me and I can still pick up the phone and call most of them and get a decent conversation. It is understood that they are under non-disclosure agreements from levels of "security" that range far up into the secret depths of national and international government affairs, so we know what we can and cannot discuss openly. They still kid me about my "novel ideas about comets" although there are a few who I know are fully aware of how correct I am about all of this (but cannot say or they too would be history). I have been told that NASA actually has a team that analyses data in terms of my theoretical work, but I can only imagine how convoluted and mixed up that group must be. Why they refuse to come to me for that function is also a mystery, but I never have claimed to be able to make sense out of why NASA does what it does.

Back in the late 1990's I was associated with a web page that was in the thick of the comet Hale-Bopp controversy. That page at the height of the battle was receiving over 3 million hits a month. Someone somewhere was interested in what was being said. My theoretical work was the core of that web page and it kept the staffs at NASA quite busy trying to keep their heads above the tidal waves of public sentiment that was generated by that page.

Much of the internet traffic which was trapped by originating addresses of the visitors to that web site saw a large volume of the traffic coming from astronomy and astrophysics departments of universities as well as continual traffic from NASA, the Department of Defense, Staff of the National Labs, the NSA, CIA and other letter agencies of the government. If comets were just silly little insignificant snowballs then it sure is unusual to see all the interest in someone saying they were not. There was also significant foreign traffic. With 3 million hits per month, the daily printout of "visitors" was quite the thing to behold.

NASA was caught lying and producing doctored photos to "prove" that the comet Hale-Bopp did not have a companion. This was a huge story in itself as they publicly accused civilians of doing the same, and causing the suicide of 39 people in what was described as a "cult-like" group called Heavens Gate. In the end even NASA admitted that Hale-Bopp had a companion, but never retracted the damage it had done to the reputations of the people involved.

It was during this time that NASA began blatantly hiding data from many space investigating facilities including the Hubble Space Telescope, the SOHO solar observatory as well as other facilities. The days of getting information out of NASA were over. They developed public relations offices whose official function was to release information to the public and to fend off problematic people like myself with planned disinformation campaigns. Congress also became off limits to inquiries into NASA. We began to call NASA "Never A Straight Answer" and the "National Astronomy Security Agency".

It was during this time (circa 1996), as the huge comet Hale-Bopp began to drive the Sun's energy to new levels on a daily basis, that many of the concepts regarding the Sun-Earth connection and Earth weather and Earth Changes became solidified, such as increased volcanism and earthquakes. Armed with a correct theoretical model, I could determine more correct information out of the limited data I could get from NASA than they could with all of their hidden data and supercomputer facilities, but using their incorrect theories.

In addition to comet Hale-Bopp, another comet named comet Lee arrived in the solar system and was the basis for my beginning to make long term weather predictions based on electrical alignments of celestial objects. This will be discussed at length later in this book. Other comets followed and the empirical data became more than proven when a series of unusual alignments with a small comet named Comet C/Linear 2001 A2-B occurred with Earth, our Moon and the other inner planets Mercury, Venus and Mars. The series of solar storms coinciding with the dates of cyclonic storms was so exact with the predicted dates that there was no longer any question of whether or not this was correct. All of this work was understood and made public as early as 1997 on the above mentioned web page (which no longer exists at the time of this writing, I now have another personal web page).

It is this work that NASA is so feverishly attempting to copy and take credit for as this book goes to print (in their recent team written book "Storms from the Sun" that I understand was written in few short months). Of course they want to give it the NASA twist (that they are the experts and to use their access to the media and press to pound their chest to the public). The problem with this book and their theoretical understandings is that they still have no clue of what really is driving the Sun and our weather. This will have devastating results if we are plagued by another large comet in the inner solar system, because they essentially are preventing the correct information to get to the public who will be left unable to respond and with insufficient time with which to prepare.

In addition to my new understanding of the connection of the electrical nature of the solar system to the Sun's behavior, and its relation to the weather on Earth, an amazing realization occurred. Since it was now understood how solar electric fields affected our ionosphere which in turn affect the surface weather, I came to the understanding of how to dampen or remove energy from massive cyclonic storms such as hurricanes and tornadoes. This information was offered for free to NASA and the NOAA (the National Oceanic and Atmospheric Administration) weather (experts) in the late 1990's but they refused to even acknowledge or respond. In the mean time, killer hurricanes such as Floyd and Mitch as well as many monster typhoons in the Pacific and Indian Oceans were devastating the lives of hundreds of millions of people.

Typhoons and cyclones coming off the Indian Ocean in one week saw over 3 million people either killed or rendered homeless. Still NASA and NOAA refused to respond or take note. NOAA still publishes their list of "factors" that cause and affect hurricanes. They claim that hurricanes derive energy from warm ocean waters, but many hurricanes exist far to the north in temperate and even icy waters. I have saved these lists of factors for posterity to prevent them from performing the "creeping crud" changes as they learn and try to incorporate the truth given in this book.

The tale of "the perfect storm" is a perfect example of massive cyclonic storms that were not caused by the "warm water" theory. (The three storms actually formed over land in winter conditions). Numerous examples exist of larger hurricanes coming in on the heals of smaller hurricanes that would have already sucked the heat out of the oceans if that concept were correct. In the discussion of how hurricanes and cyclonic storms really work it will also be explained why the warm water theory does not make any sense.

It would be like someone seeing a car drive down the street for the first time and when asked what propels the car (and without knowing there is a powerful engine and gasoline doing the work), the person looked at the outward physical view and stated that it must be the exhaust coming out the exhaust pipe that is pushing the car down the road. Simple energy calculations alone will prove that the warm water theory cannot account for the massive energy found in hurricanes. It is interesting to note that the planet Mars has been observed to develop huge hurricanes yet there is NO water there to create or sustain them.. Clearly some other process is at work.

The reality is that all of these storms are caused by powerful electrical connections between the Earth's cloud systems and the electrically charged ionosphere, which dramatically increases in response to solar flare conditions. This is how the currents induced by solar flares find their way into our atmosphere to excite and drive these powerful and dangerous storms. The entire process for how this works has been worked out and is explained in the chapter in this book "The New Solar Model and Space Weather". Solar electrical conditions do affect our weather in a very direct way. What can be stated here additionally is that comets can and do affect our weather, as amply described by the ancients, and we are now seeing and proving this with exact scientific measurements today.

The important aspect relative to this book is that comets can affect our weather and the very large comets can have devastating affects on our weather in addition to the many other disastrous effects that can be caused by comets. This can happen when the comet is not even near Earth (action at a distance). The many factors that can cause this are explained in this book. You will notice the newly released book by NASA on the Storms from the Sun laughs at the possibility of comets having any affect on Earth or its weather. They are very wrong.

The Hale-Bopp days saw enormous tropical storms with wind speeds over 300 miles per hour and monster tornadoes that were over a mile wide at the base and running for up to 100 miles on the ground. Much of the devastation of these storms could have been averted had NASA and NOAA responded to this free offer to help direct resources to build the simple systems required to remove energy from these massive storms. In spite of hundreds of meetings each year in the officialdom of these scientific communities, these groups still have no clue what really causes the weather. They are misdirecting the vast publicly funded resources for their own fame.

This has prompted me to include later in the book an article derived from the Hale-Bopp and comet Lee era which was directed to NASA and NOAA entitled "You Can't Predict the Weather if You Don't Know What Causes the Weather … or why are school children in Florida still in school as a CAT 5 hurricane (hurricane Floyd) approaches off shore?". What makes this story even more interesting is that I had made a prediction 5 months earlier that comet Lee (due to peculiar electrical alignments with the planets and our Moon) would on that precise time frame have a high probability of exciting multiple large hurricanes in the Atlantic Ocean. At the time hurricane Floyd reached Florida, there were at least 4 named hurricanes in the Atlantic with one bearing down on the heals of Floyd (something hurricanes are not supposed to do in the warm water theory).

The Plasma Discharge Comet Model also explained why hurricanes dissipate as they encounter land (and for example why the Great Red Spot of Jupiter can go around and around for centuries without stopping … there are no continents to short out its electrical circuit). Hurricanes dissipate power when they arrive at land because the ocean is a poor conductor and the vertical currents have no immediate connection point to "short out". As the hurricanes approach land the electrical currents that are driving the storms short circuit to ground and the storm begins to break up and drop the water it has accumulated while at sea. The entire formula for how these storms are powered from the charge in the ionosphere is beyond the scope of this book, but there is no longer any question regarding its correctness.

This process then became the basis for shunting these storms by the same process before they arrived at land and caused their destruction. Essentially there would be a number of ways to "short circuit" the storm while still out at sea and at least reduce the affects as it approached land. This would also cause the storm to dump its water load at sea and before the massive flooding moved over land. Hurricane Floyd for example flooded the entire eastern United States causing damage that still is not repaired today. To date NASA and NOAA maintain their incorrect course and these states remain vulnerable as much today as always. Future storms could be reduced but it will take resources and the inclusion of the people such as myself who now know how to do this and can convert theory into practical processes.

One last item is that at the beginning of the 1900's a man named Tesla learned that there was a vertical electric field near the surface of the Earth and he learned how to tap into this to derive electrical power. What the Plasma Discharge Comet Theory now does is to explain where that energy came from and how to tap it as an endless free source of power for Earth. When one imagines that the energy in one huge thunder storm is enough to power our entire country for a year, it no longer makes any sense polluting the world with fossil and nuclear fuels that require burning to produce energy.

The last point to mention here is that recently NASA made a public press release that its scientists thought that the Sun would now start to calm down and not to worry. The solar maximum was supposed to have taken place in the year 2000. By now we should be well on the road to a solar minimum, yet the Sun never subsided in energy or flare production and is increasing to record levels on a daily basis as this book goes to press. There is sufficient evidence based on the current solar activity level that indicates that there must be a large object breaking into the far reaches of the solar capacitor and is already affecting the Sun. If this is the case and this continues to come closer into our part of the solar system there could be devastating weather on Earth. If this is not the case the reader can rest assured that eventually a very large comet will again enter the inner solar system. Until we are prepared to meet it with proper knowledge and resources to sustain the "siege", the world population will be at risk of repeating what we know has happened in the past when these devastating events occur.

Many people admire Carl Sagan, probably the most famous NASA scientist. My personal opinion is that I did not like the attitude that has been presented … that NASA scientists are brilliant above the rest, and they will bequeath upon the average human being the "results" of their studies, but the background reasoning is too complex, "so just believe us". Few people realized the number of scientists who objected in the background but with no access to the news media or public. This is as much the fault of the news media as these scientists' arrogance and control.

One last point regarding NASA is that it has long since become "Congress-proof". This is a term coined to describe the process of distributing the NASA government contracts around the country to every jurisdiction. The original reason was to make sure that many small businesses had access to government funding that typically only went to the large aerospace firms. The end result however has been that this has now been turned around in to a form of congressional blackmail. If any senator or representative voices objection to the operations of NASA, then his district will be cut from the government funding that flows there.

This is the exact same form of blackmail that befalls news reporters who do not follow the NASA party line and report only the good side of NASA … and … no pointed or problematic questions please! … or you and your publication will never get another story from NASA. Additionally, do not ever think of publishing a story from those outsider "pseudo-scientists" because the same fate will befall you. Essentially it is professional suicide to break this protocol.

The NASA news release system is strict and comes only from designated NASA news distribution points in Goddard and JPL. Individual scientists are under strict non-disclosure agreements not to disclose anything individually. All news outlets must go through designated channels. These scientists are equally restricted under the NSA decree that they cannot discuss or admit publicly to any event or situation that might cause public alarm. That is why the data regarding a new arrival like Planet X will not be allowed from the halls of NASA. They are under strict contract ***not*** to tell the public.

VI. COMPARING TRADITONAL ASTRONOMICAL THEORY (NASA) WITH THE MCCANNEY PLASMA DISCHARGE COMET MODEL

This chapter will attempt to simplify the comparison of the two "camps" of thought being discussed in this book. As you will see they are not compatible and in fact say exactly the opposite on many details. Yet they both propose to explain the same reality and data that has been obtained from space probes and other modern equipment.

What I learned to look at early in my mathematical and physics career were three types of tests. One was the "contradiction test" that says 'if a single contradiction occurs then one must pay more attention to that than all the other data that does fit the model'. Unfortunately what the modern astronomical community has allowed itself to do too many times, is to "sweep the data under the rug". This refers to those pesky little contradictions of which I am speaking.

The second test I call the "extremes test". That is you take a concept to an extreme condition and this many times allows you to decide between two concepts or explanations that otherwise seem to give similar results under normal conditions.

The third I call the "devils test". This refers to basically being your own worst enemy. You have to be the little voice that opposes and temps you as you try to decide what is right or wrong. In doing this you many times reach the numerous objections that would be raised by opponents. If you fail to pressure yourself or your respected piers, then it may all come home to roost as is now occurring in this book ... that is, it became too easy to not apply the devils test within NASA when and where it was needed.

In applying these three tests to standard astronomical theories of the solar system, I found that they failed repeatedly on all three types of tests. This is what led me to question and then replace the basic axioms. It also helped to have had a storehouse of the work already completed so that the visions of what were needed were already in place. It also was invaluable to be at the very data center where all the most recent data was pouring in, and to see on a daily basis how the standard explanations had to be abandoned or altered to make the data fit … or as in many cases simply ignored by NASA "until proper explanations were found".

The following list is intentionally abbreviated as it could run for dozens of pages. The intent is to contrast the basic differences between the two ways of viewing the solar system. It would be good to note that the NASA theories have continually been modified over the past decades to almost become exactly like the fundamental results of the Plasma Discharge Comet Model, whereas this later concept remains intact in its original unaltered state. What is being found today in space probe data was anticipated and predicted accurately by this model over 20 years ago.

An outline format will be used listing the main topic, followed by a) the NASA traditional explanation and b) the Plasma Discharge Comet Model explanation. Although this may seem a bit tedious and detailed, it will be helpful in understanding the explanations in the chapters on Earth Changes, etc.

I) Origin of the solar system

a) NASA – the solar system formed all at one time 4.5 billion years ago and no new planets or moons have formed since

b) Plasma Discharge Comet Model – Only Jupiter and the Sun were in the original solar system and all the other planets and moons were captured and adapted into the solar system one by one at later times. Each new object formed under unique conditions and were large comets (the cosmic vacuum cleaner), until they settled into their final resting orbits. As new members arrive they will encounter the currently existing planets and moons in a dynamic game of celestial billiards.

II) The circular orbits of the planets

a) NASA – the orbits formed in nice circular patterns at the time of formation and the planets and moons have not altered from these since.

b) Plasma Discharge Comet Model – as the comets are drawn into the solar system and are captured and encounter other planets, two effects tend to reduce the comet orbit and bring it into a circular orbit. First it will encounter other planets and either gain or lose energy. The primary effect that will reduce the orbit of the comet (and new planetary member) is the "tail drag" caused by the pull of the tail as the comet draws in vast amounts of matter relative to its size. The huge comet Hale-Bopp had its orbit reduced from 4200 years to 2650 years in one passage of the Sun. The dirty snowball "jetting concept" could not account for this amazing change in orbit.

III) Comets

a) NASA – comets are tiny "dirty snow balls" that were formed at approximately the same time as the rest of the planets. Since standard theory requires a huge solar wind emanating from the Sun to account for the spin rate of the sun, these comets somehow must have been moved out to a location where we cannot detect them, called the Oort cloud. Occasionally objects pass by the Oort cloud and disturb the comets from their orbits and they come in towards our Sun. These snowballs melt (sublimate) when near the Sun and the solar wind sweeps their tails out away from the Sun. Comets are getting smaller and go away eventually or leave a small asteroid in its orbit. NASA states that only if one of these tiny nuclei hits Earth directly can it harm Earth.

b) Plasma Discharge Comet Model – Comets are asteroidal or larger rocky objects, which ignite a complex electrical discharge of the Solar Capacitor when they enter the Sun's electrical region (that is produced and maintained by an excess current of protons in the solar wind). Comets are *drawing in the tail material* and therefore are getting larger as they move on their orbits in the solar system. Although most comets are small, some can be large or even of planetary size, and if so, can cause extreme damage to the existing planets and moons as they work their way into their eventual stable orbits. They do not have to directly hit an existing planet to cause damage, in fact extreme levels of damage can be caused by a myriad of effects by "action at a distance".

IV) Venus

a) NASA – Venus is an original planet of the solar system and was originally expected to be Earth's "twin". However, when the high temperature (hot enough to melt lead)and extensive atmosphere (900 times as dense as Earth's) were discovered, the Green House Effect was invented to account for this. There is no explanation for Venus having over 1 million volcanic calderas on its surface (e.g. it is venting heat from the young extremely hot planetary interior).

b) Plasma Discharge Comet Model – Venus is a new planet and is exactly what you would expect from the recent acquisition and formation from a large violent comet. Since its orbit lies inside that of Earth, it had to have encountered Earth on multiple occasions as it nestled into its current orbit. It will now begin to cool down and eventually will become a water planet as it will drag in water from the solar system as Earth is doing on an ongoing basis today.

V) The Sun

a) NASA – the Sun has a nuclear fusion engine in its core (ignited originally at the birth of the solar system by gravitational pressure) that heats it into a hot glowing ball 6000 degrees and it will burn for billions of years at a steady rate. The Sun is electrically neutral and is not affected by external conditions (especially not comets which are insignificant little snow balls).

b) Plasma Discharge Comet Model – the sun's nuclear fusion is taking place in the upper atmosphere and the nuclear fusion reaction is ignited on an ongoing basis by pinched electrical discharges in its violent stormy atmosphere. The Sun's temperature actually cools as one measures deeper, eventually hitting a solid planetary type core (this is confirmed by x-ray measurement experiments performed by Russian scientists). The Sun is dynamic and changes rapidly to external conditions. Especially, comets which are forming an electrical connection with the Sun can rapidly and significantly alter the solar output; large solar flares can occur in response to electrical interactions with comets. Additionally special conditions such as electrical alignments of the planets can occur which will cause a discharge of the solar capacitor which in turn can cause currents to flow to the Sun, resulting in enhanced solar flaring and energy output

VI) The electrical nature of the Sun

a) NASA – the Sun is electrically neutral and so is the entire solar system. Energetic effects are explained to be results of magnetic effects such as "magnetic reconnection", magnetic pick-up ions, magnetic bubbles releasing energy, and magnetic storms on the Sun.

b) Plasma Discharge Comet Model – the Sun is left with a net negative charge as it continually ejects large composite streamers of primarily protons in the solar wind. This builds up and sustains a "solar capacitor" with the positive anode far beyond the orbit of Pluto, where there is a large ionized cloud of dust and gasses. This capacitor is discharged by the asteroids and larger comet nuclei that enter the solar system. This capacitor is what powers and controls the electrical nature of the solar system. Magnetic effects are secondary effects of the electrical plasma currents that flow as a result of this power source. The Sun produces thousands of times more energy in the form of electrical energy as it does in the form of light energy, but Earth is buffered from this by its large magnetic field. We only notice changes here when the local electric field and current flows are altered either by solar flares or large comets passing nearby. The planets in near circular orbits also constantly discharge the solar capacitor, but to a lesser degree than comets which are on elliptical orbits which cut at greater angles through the solar capacitor (causing a greater degree of discharge).

As one might imagine, this list could go on for pages, but this should be sufficient to give the idea of contrast. The two approaches are almost diametrically opposed on almost every issue, yet they both claim to explain the same phenomenon. When giving the three tests mentioned before the outline, it will eventually be clear which set of principles is correct. There is scant few ways to combine them to form a common ground.

If we are dealing with science, then experimental results should be the only deciding factors, not popular vote of the referees of the journals. This author has listed at the inception, in the original papers, a list of experiments that will use the above differences and through rigorous experiments will prove or disprove one set of concepts or the other. These experiments have been on the table for decades but to date those with the publicly funded resources have chosen to ignore them and depend of the popular vote of their piers to keep the current

Axioms of Astronomy in its hallowed place. This clearly shows that we are not dealing with *science* anymore.

As an example, one of the primary predictions of the Plasma Discharge Comet Model, was that comets would be found to produce x-rays to the sunward side of the nucleus. Whereas the Dirty Snowball Comet Model clearly did not make any such prediction, nor could it explain such a phenomenon by any rational means. The prediction that the x-rays would be found out away from the nucleus was even more remote. This prediction was the most telling of all since one would never expect the docile little snow ball to produce x-rays that could be visible and measurable from Earth, let alone far out to the sunward side of the nucleus where the Dirty Snowball Model only saw the same solar wind that existed everywhere else in the solar system.

The prediction of x-rays of course was predicated on the existance of the high intensity and high energy electron beam coming out from the Sun that is an essential but otherwise invisible characteristic of the Plasma Discharge Comet Model. There were other distinctive features of this model which predicted other effects that could be measured. So the entire issue should have been decided, as science should do, with experimental evidence.

In 1985 the announcement was made that a spent satellite that had been monitoring Earth's outer magnetic field called ISEE-3 would be redirected to rendezvous with comet Giacobini-Zinner. This was a huge coup since NASA was not sending a space craft to Comet Halley. In anticipation of going to a comet that they expected to be a snow or ice-ball, NASA renamed it ICE-III. The team leader Dr James Farquhar was turning a dying satellite into an international best seller.

I was familiar with this satellite and saw the golden opportunity for my x-ray hypothesis to be tested and verified. To my horror one day I heard that to conserve electrical power on the flight, the x-ray experimental detector would be shut down. I called Dr Farquhar and explained the situation. He consulted with the "comet experts" at NASA and the decision came back that there was no need to leave the x-ray detector on since there would not be any x-rays around the snowy nucleus of this small comet.

I then called and spoke with the two PI's (Principle Investigators) on the x-ray experiment. They were totally in shock also. Here was a completely unexpected turn of events for a satellite that had previously been a discrete no-name project that only a few plasma physicists even knew about, and it was now turning into a huge international party … and they had just been dis-invited! I explained my theory and the prediction. They were not sure how to deal with this, but with all the bidding and coaxing, the x-ray detectors remained off throughout the entire ICE-III mission to comet Giacobini-Zinner.

I also told them to pass to the sunward side of the comet to measure the electron beam as a second way to confirm my theory and disprove the Dirty Snowball Model. They ignored that request also. The encounter and data that resulted brought the expected series of "surprises", but the flurry of papers that followed by the NASA scientists took care of any discrepancies. Once again I could see the signatures of the Plasma Discharge Model, but you cannot talk to NASA scientists.

The entire issue of x-rays in comets lay dormant for a decade until one day a friend of mine called me in 1996 and said, "did you hear that NASA just announced they had discovered x-rays to the sunward side of a comet nucleus". I got the news release which said "the discovery of a strong radiation signal- about 100 times brighter than even the most optimistic predictions, was made March 27, 1996…". I immediately called all the scientists who were working in the Goddard Space Flight Center group. The response was universal, they had no clue that there could be x-rays from comets and it was a total mystery. I Next-Day-Air mailed my papers to that group of scientists.

Since I knew some people inside Goddard, I got a first hand account of how my papers were received (since it meant a challenge to the Dirty Snowball Comet Model). My papers were never mentioned in any press releases, were never referenced in any papers written by NASA scientists, and when I went to the international scientific meetings regarding this discovery, I was not allowed a speaker slot to present the prediction.

When the news conferences were held I was not told when or where they would be. The only way I found out was when a reporter told me and when I went to the news conference room it was locked with the NASA scientists inside. It was a totally controlled affair. I also heard from an inside contact that the leader of the Goddard group told the young physicist who actually made the measurements, that he should not read my papers. When I later read that leader's paper of "his explanation" for the x-rays, it had essentially my ideas but tacked onto the Dirty Snowball Comet Model as a basis.

In the same press release the NASA spokesperson commented, “We had no clear expectation that comets shine in x-rays,” ... “now we have our work cut out for us in explaining these data”.

This discovery turned out to be one of the top 10 discoveries of 1996 according to DISCOVER magazine, almost 20 years from my original prediction date. The DISCOVER article touted the NASA discovery and of course never mentioned my work.

As one should expect, NASA scientists now have numerous theories explaining how the x-rays form from a snowball. Recently an article stated that NASA scientists have self-confirmed one theory as the cause. I have since reviewed the “confirmed theory” and found it fails to explain the level of x-rays, the multiple types of x-rays and it fails to explain what the news release called “pronounced increases and decreases in brightness ... typically over a time difference of a few hours”. Scientists of course might argue over these fine details, but what is clear is that only certain people are allowed into the chamber to present their results and arguments. This is a perfect example of what I call “the creeping crud”, that is, the continual assimilation of unexpected results into the Axioms of Astronomy . It is the standard way NASA has operated since the Space Program began.

I have hundreds of stories like this in dealing over the decades with NASA and the astronomical community from the outside. But the most telling I will include here because 1) it is well documented and 2) it is so blatant that the reader will get the idea of what “suppression of intellectual data” is all about ... NASA style.

In the early 1980’s the Hubble Space Telescope” (HST) was under construction (another NASA fiasco in its own right). A year or so prior to launch, as a public promotion and to appear to be open to the amateur astronomer, NASA promoted a contest to allow amateurs to create observational programs and compete for the amateurs dream ... a slot of time to use the HST.

I of course qualified as an amateur, since NASA had itself taken care of the fact that I would never make a living or get any grants, etc. as an astronomer. The competition was quite intense and I was surprised at the level of detail and minimum competence that one would need to even enter this contest.

I sent for the information and decided to see if I could beat NASA at its own game. The applicants had to integrally know the workings of the systems of the HST and understand quite complex information regarding filters, the optics, pointing hardware, slewing rates, temperature factors and on and on. Like I said, you had to become a mini-expert in the construction details and operation of the HST. Additionally you then had to construct an experiment and have it in a format that complied with a professional astronomer's expectations, since teams of NASA astronomers were to be the judges.

I was not that familiar with large telescope jargon or operations so this was as much Greek to me as the other amateurs. In fact with the level of amateurs around, I felt I was at a definite disadvantage. This was a real challenge, but it also solidified what I very much needed ... that was to define for NASA the experiments that would prove my theories correct and disprove their theories. I already had my list of experiments to perform based on theory, and this was the ideal chance to define it in exactly the terms that NASA needed to perform them (exactly as I would specify and without their meddling).

In the Amateur HST Contest there were 8 categories in which to enter. You could enter as many as you wanted, but entering even two categories would be almost impossible I was told. Armed with my already defined experiments, I entered five of the categories.

Each category then had an independent team of NASA specialists that would take the applications through an initial screening to eliminate the first round based on technical merit and to verify that the applicant had followed all the proper procedures, formats and showed understanding of the HST. After this, there was a second round, which would reduce the number of applicants that would enter a third and final cut. At this point there would be 30 finalists who would be judged by a composite panel, which was comprised of the judges of the original 8 groups, and headed by a NASA top figure. Of the 30 finalists, only 5 would be selected to actually get to perform their experiment on the HST.

Originally there were over 250 contestants. At each cut I received the letters of acceptance or rejection. One of my proposals was cut on the first round, with 4 remaining to go into the 2nd round of judging. Much to my surprise, all four of these made it through the next round and ultimately all four were approved to go to the final round. I had four of the final 30 proposals. At this point I talked with one of the scientists and he stated that they were surprised that my proposal contained not only experiments, but original theoretical work also. None of the other proposals had any theoretical work.

Apparently there was a long stretch of silence at the meeting on the final day of judging when the 8 separate groups, all of whom performed their judging completely separate from the others, reported their selected proposals for a total of 30 finalists. I know that one of my proposals came in "first" in its group, and it was hinted that actually all 4 were the number one recommended proposals in each of the four groups that I entered, although I could not confirm that independently. Now one has to ask what the probability of the end results could be. The lead scientist took the final 30 applications under advisement, and as you may have already guessed, of the final 5 who would get to use the HST, not one of my 4 proposals was selected.

Of course I did not stop there, but found a way to propose to NASA that they should perform my experiments using the unique capabilities of the HST. They have never performed any of them to date in the manner required to disprove their beloved theories. It is interesting to note that any one of the proposals, if successfully performed on the HST, would put a serious dent in the side of the traditional astronomical ship. I did this carefully when writing the proposals since my wildest hope was to get only one proposed experiment even close to a finalist position in the amateur contest.

VII. VISIBLE EFFECTS & EARTH CHANGES CAUSED WHEN NEW COMETS AND PLANETS ARRIVE

"It is not the strongest of the species that survives, nor the most intelligent. It is the one that is most adaptable to change."
......Charles Darwin

The rest of this book will concentrate on the scientific explanations of what one might see and experience if (and when) a large new planet is adopted into the solar system. Remember that a planetary type of object coming into the solar system on an elongated elliptical orbit will ignite the solar capacitor and become a huge comet. So you will not see a big red ball floating through space for example. You will see possibly, as in the cover Indian rock painting, an occasional glimpse of the nucleus through the comet coma or tail. The Plasma Discharge Comet Model explains that as the orbits of comets become circular due to the "tail drag" effect, the discharge of the solar capacitor becomes less. Therefore the comet tail becomes less pronounced. At any rate, all of the planets are still involved in minor ongoing discharges of this solar capacitor and in fact have comet like tails ... misinterpreted by astronomers to be "magnetic tails".

A comet's size depends on many factors. These include the angle it comes in on relative to the solar capacitor's structure (called electrical field lines). It depends on the velocity (or how quickly it is discharging the solar capacitor) as well as the size of the comet nucleus (since a large nucleus can maintain a larger surface charge and even an atmosphere, and therefore a larger electric charge than a smaller comet nucleus). It also depends on how close or far away it is from the Sun and whether it is moving in one of the Sun's known electrical current sheets, whether it interacts with a solar flare and so on. With so many factors, clearly this is not a simple system.

As the comet charging varies as it moves in the solar system, so too will the amount of material it is dragging in from the tail region. The amount of this material that will be retained depends on the comet nucleus' overall gravitational field. These factors determine how much its orbit will change due to this "tail drag", as the comet nucleus becomes larger.

There are many factors that affect how the comet will develop and the overall properties of the comet itself. If it pumps up the Sun into higher proton wind output or flaring, then they will cause each other to increase in size and energy. The comet is now moving amongst the rest of the members of the solar system and may encounter them or affect them in numerous ways. There is the very remote possibility that the comet might crash directly into a solar system member as did comet Shoemacher-Levy 9 with Jupiter. Our Moon is certainly testimony of millions of such direct collisions over its history, although there is a major discrepancy between the number of craters on the near and far sides of our Moon. The near side "seas" or "Mares" are clear evidence of very recent large scale volcanic activity that was caused during the last major comet encounter, as noted by Biblical reference when the "Moon turned to blood".

But the majority of affects that a new planet in the "comet stage" will cause on existing members of the solar system are what I call "action at a distance". The point is that the comet does not have to be anywhere near the planet or moon to have sometimes dramatic affects on it. In addition, once the planets become involved say for example in a three way electrical alignment or attachment, greater affects may be unleashed from the energy stored in the solar capacitor. The key to understanding is that the Sun's capacitor, which has been described frequently in this book, stores unimaginable amounts of static electric charge and energy. When a comet or series of planets, moons and comets become involved, then the level of power released from the capacitor increases and that power usually is passed along the paths that include these planetary objects. These are the "electro-magnetic plasma" affects. The next 4 chapters of this book will go into detail regarding these effects.

In addition to the electro-magnetic plasma effects, there are the physical or what one might call Newtonian gravitational effects. These would involve the alterations of orbits due to gravity of one object working on another, or if two massive planetary sized objects interact at close range, the "tidal waves" can be set up in the outer structures of these planetary mantles. These effects can range from small ocean tides such as our Moon causes every day, to massive major Earth altering events such as mountain building, earth quakes or creation and eruption of volcanoes.

All of these then have secondary effects that may last days, months or centuries. For example if a large comet nucleus passes near a planet like Earth, causing continental plates to shift and compress, these may then continue to "relax" for centuries to come and long after the memory of the comet passage had faded from human memory. A future chapter will detail how the concepts of "global warming', the "melting glaciers", the observed movement of tectonic plates or "plate tectonics" are NOT caused by the effects commonly attributed to these such as greenhouse gases, or meandering tectonic plates. These are in fact the Earth relaxing from its normal state after the last large catastrophe of a few thousand years ago. The Earth's population and history are also recovering from that event. There simply has not been enough time in the standard explanation of man's evolution for the various races to have evolved, especially now that we know there were races living in South America long before the Egyptians occupied ancient Egypt.

If earthquakes or volcanoes occur under the ocean, then huge tsunamis (induced tidal waves) may occur and continue to occur as these conditions last for centuries after the comet encounter. Underwater volcanoes may cause water to boil and have drastic affects on sea life, or cause species to adapt or die off. Mass extinctions and stressing of biological life systems would be one secondary result.

Another result might be that the Earth or another planet would be engulfed for a period of many days or longer in the coma or tail of a very large comet. In 1910 Earth passed through the tail of comet Halley. There was tremendous scare and people committed suicide and there was general panic due to the belief that poisonous gasses would kill off large populations. Nothing happened so astronomers love to use this as an example of comet hysteria gone awry. Halleys however is a small comet. Comet Hale-Bopp was a large comet and Earth came close to being caught in that comet's huge coma for a number of days or weeks. This would have been disastrous, and will be discussed in the following chapter dealing with Hale-Bopp as an example of a comet that could have caused severe damage to Earth if conditions were only slightly different. If there was no harm in comet Hale-Bopp, why did NASA hide its data and photo sets on that comet, and why are they lying about the size of the nucleus? As it turns out, the nucleus of Hale-Bopp is about the size of Earth's Moon.

Yet another effect of a large comet could be the interaction of that object with the existing atmosphere of a planet. We have two perfect examples that were recorded in man's recent history and we see the results of this today evidenced in the wake of these events. One is seen in the planet Mars. Its atmosphere and oceans were removed recently. The ancients are said to have witnessed this and described it as a giant snake-like extension coming out of the huge comet that eventually became the planet Venus. Based on the Plasma Discharge Comet Model, I will describe the process and what caused the appearance of a huge cosmic snake attacking Mars in the upcoming chapter. I believe, as verified by the ancient Egyptian priest cults, that Earth has experienced multiple such events in the past 10,000 years. Some were more devastating than others.

During the same passage of that huge comet that became the planet Venus as recounted by Velikovsky and the other events attributed to the giant object called Nibiru, Earth received what I call "extreme pollution events". This included large influx of water (flooding) as well as tidal wave flooding of sea shore regions. There were periods of brief mountain building, massive electrical discharges between comet and Earth (e.g. the burning pillar of fire witnessed by Mosses as they approached the Red Sea during the Exodus that sounded as incredible explosions marching down the Sinai).

The burning "fire and brimstone" was the meteor streams and influx of hydrocarbons from the huge comet tail that polluted the Earth and gave us the extensive oil fields that cover the Earth and line the bottom of the seas and oceans. The La Brea Tar Pits are a prime example of this pollution event, and also a prime example of how ridiculous the convoluted stories can become in trying to explain away the obvious. The standard stories say that basically thousands of these animals were so stupid (including at least one human) that they either wandered into the tar pits or chased one-another into it until they were completely submerged and died.

What really happened was that the pollution event fell over the entire planet and the animals were literally drowned in a rain of burning oil from above. They died, and as the liquid oil sank into the Earth, the tar pits remained with the dead animals entombed in the thicker tar. The entire issue of where oil comes from and its location on Earth will be discussed again.

We do not have to fight a war in the mid-east for oil, as it is found the world over and in abundance. It is not the result of millions of eons of decaying plant life. If it were, where were the fern forests that grew under the oceans?! I have yet to see any chemist show how dirt decays and transforms into crude oil and natural gas. There are literally oceans of oil all over the Earth. It ARRIVED ON Earth as part of multiple large-scale pollution events from passing large comets. It is not a rare or limited commodity, and it is not the rare substance controlled by just a few elite people. There are better ways of powering our need for energy (e.g. tapping into the solar electric field as Tesla was doing for free). The real reason we should cease using oil is because it pollutes the environment.

Coal is formed from compressed plant life that was rapidly mounted into huge piles and DID NOT DECAY SLOWLY. It formed as the tidal waves and tremendous winds raked up plant life and piled it into heaps in few short days of massive devastation. A huge storm that hit Madagascar a few years ago cleared out huge sections of forest and piled it in this manner, so it is clear that the massive wind storms that would accompany the passage of a huge comet would perform this act on a world wide scale.

The Hopi Indian legends describe the "Devil Winds" in the 7th and final level of Earth Changes. They discuss this destruction and plight of man during these times. They go on to say specifically that the final stages are caused by comets, and that "the blue ones are the dangerous ones". We now can interpret this ... as the upcoming chapter on Earth weather will explain how electrical fields and electrical currents from the Sun and comets can drive cyclonic storms and develop straight line winds on Earth. The "blue ones" refers to comets that have such an intense electrical discharge passing though them (the electron current coming out of the sun), that they literally light up like a light bulb.

When Mosses witnessed the burning cyclone of fire extending into the sky, this was literally the electrical discharge from the passing new planet (the comet) and Earth with the burning naphtha pouring in. That anyone survived at all is most remarkable and a tribute to their tenacity. They wandered for 40 years in the "desert" left in the wake of the comet's devastation, not because they were lost, but because there was literally no place to go. Everything had been destroyed.

They speak of the rumbling Earth and constant earthquakes, of boiling water and noxious fumes belching out of the Earth for years after the event. Unlike the Hollywood movie, most of the people died. When the Red Sea parted, the shifting Earth moved and threw the waters on the approaching Egyptian army.

There are many misconceptions regarding the extent and cause of pole shifts. One common statement says that the magnetic field of a new planet will interact with the magnetic field of Earth like a bar magnet and "flip" Earth. There are a number of incorrect assumptions in this description. Only in the Earth's very central core is there a real permanent type magnet. The other "layers" that make up the magnetic field are in loosely held pockets and regions of the Earth's crust and mantle. The rest of the field is in the form of electrical currents in the Earth and in the ionosphere and above. Additionally, if the Earth receives a magnetic "jolt", the Earth is not a solid piece of material as a bar magnet you put in your pocket. The individual regions would react and cause increased volcanic activity and earthquakes, but this would not cause a pole shift. Even if the Earth were a solid magnet, the magnetic jolts would not be enough to move the Earth or cause a pole shift. One geologist recently suggested that the Earth could do this kind of pole shift on its own, whereas another suggested that a lopsided melting of polar ice caps would cause wobbling and a pole shift.

There is only one process that has all the factors sufficient to cause a shift or rotation of the spin axis of the Earth's mantle, and that is a gravitational wave set up by a passing large comet. The "tidal wave" is set up in the Earth's mantle just as an ocean tide occurs in the Ocean. Our Moon twice daily draws the mantle up a few inches as the Earth spins under the Moon. This goes unnoticed, however, a large object passing at high speed would set up an incredible tidal wave in the Earth. This has happened and is the cause of dramatic Earth Changes up to and including mountain building.

Our mountains were not built by the slow moving of tectonic plates. There are a number of reasons why this slow process simply does not work. The first is that the Earth's mantle is not rigid. So forces created thousands of miles away would be absorbed locally and would not force the billions of tons of rock skyward for miles as is seen. The mountain chains were ripped from the ground in a brief and violent motion as a massive planet/comet passed by Earth ... sending its tidal wave racing through the mantle at thousands of miles per hour.

Another basic theoretical reason why plate tectonics does not work is also related to the fact that like many space scientists and astronomers, most geologists are not well schooled in physics. As a result, no one has ever defined or determined how the Herculean transverse (horizontal) forces, that supposedly move entire continents around the globe, are derived from ***vertical forces*** that occur thousands of miles away. As with the theoretical problems with the Greenhouse Effect, nicely drawn pictures do not answer these fundamental questions. This problem with the tectonic forces I call "the hammock" problem. With a hammock that is stretched tight between two posts, say for example that a 100-pound person lies in the hammock. The vertical weight supported is 100 pounds. This, however, translates into thousands of pounds of tension horizontally onto the ropes of the hammock and the poles supporting it.

The problem is somewhat in reverse with tectonic plates but the principle is the same. The forces inside the Earth are due to heating and expansion and therefore translate into vertical forces which sometimes break through and form volcanoes locally. Now apply the "hammock principle". To translate this vertical force into a horizontal force requires a multiplication factor of thousands of times the required force. If there were this level of vertical force then it would simply break the surface where it existed and form volcanoes in that point. Additionally, as already mentioned, if some did translate horizontally, it would compress and deform the continental shelf locally and would not translate to mountain building in the middle of a continent thousands of miles away.

The Earth is much like an egg in that the mantle is not connected rigidly with the inner core. Thus when a tidal wave is set up in the mantle, it will move over the core. The old north pole of Earth was at one time just north of the state of Wisconsin (what we call the Laurentian Ice Sheet or glacier was really the old north polar cap), and the old south pole was in the south Indian Ocean. At that time Russian Siberia was a temperate climate with herds of tropical mastodons and forests of tropical plants. The current north polar cap was open ocean. The ice covered land mass we call Antarctica was a temperate continent with wooly mammoths and possibly human populations. This was in the pre ice-age era of Atlantis and Mu.

This chapter has been an introduction and running summary of the effects that would be experienced by a large passing comet. There are literally hundreds of other subtle effects and "aftershocks" that would occur such as extended auroras, a possible breakdown and/or reversal of the non-permanent magnetic field, etc. These in turn would have other secondary effects such as increased solar radiation that would increase cancer rates and even possibly accelerate genetic change and biological evolution. This book is just an introduction and could not possibly list or elaborate on the extensive changes one might see.

One key question that I am asked frequently is where would be a good place to go and how or what would one do to survive. These are difficult questions and I have heard all kinds of bizarre answers such as the existance of "energy vortices" that will protect certain areas. The simple answer is that there is no really safe place because of the myriad of effects. Where you would be safe from one effect, another might endanger you. And even if you survived the initial devastation, one would then have the task of longer-term survival in the face of great odds. Most people today would have a hard time surviving for a week without the nearest grocery store or local gas station open.

Some ask if they should buy a generator or other equipment. In the face of severe devastation from multiple effects, the conditions would be so difficult that you may not be able to navigate or move more than a few hundred yards let alone move across a larger area.

This is not meant to scare people or instill fear into the public. This is simply saying what others will not tell you … that is the truth about the level of severe conditions that would accompany a passage of a large object through the inner solar system. As mentioned previously, it does not necessarily have to hit Earth or even pass that close to Earth to have severe effects. It is clear that the scientific community is preparing to minimize the approach and effects of any new objects as they have done in the past. Some even think they are correct. They cannot conceive of such events with their incorrect concepts of how the solar system functions.

The following Chapters will discuss some of the Earth Changing effects in more detail and will discuss the cause and effect relationship between the Sun and Earth and other objects such as new large comets.

VIII. THE NEW SOLAR MODEL AND SPACE WEATHER

This will be one of the most copied chapters in this book as the space science community is slowly stumbling onto the realization that space is rampant with electrical currents and powerful forces. Every international scientific conference today is filled with questions that are probing the sources of these effects. This will be the first exposure many of these scientists will have to my work, so this will be as detailed as possible to give the overall workings of what is commonly called "The Sun-Earth Connection".

One cannot understand this under the guise of what is called "Chapman Physics". Decades ago, Chapman defined a simple problem and with it "proved that outer space had to be electrically neutral". The only problem is that nature did not know about Chapman or his rule and in fact has a different set of rules. Chapman's problem went something like this. Imagine a region of space with an object in that space and, for whatever reason, the object became electrically charged. You can easily calculate that the plasma (rarefied gas clouds that permeate the solar system in which electrons roam free from ionized atoms and molecules) would very quickly move to cancel out the charged object and shield it from the rest.

Every student of astronomy and astrophysics learns this in one form or another, thusly convincing them of the electrically neutral nature of the solar system. When I developed my theoretical work, I realized that there was in fact plenty of evidence that there were huge electric fields in space that caused huge electrical currents to flow. These in turn created the magnetic fields that were being measured all over the solar system. While space scientists were developing "magnetic" explanations, I was probing one level beyond for the root cause of the effects I was seeing in new space probe data that arrived daily at the Cornell University Space Science Department.

My conclusion was based on the realization that comet nuclei had to be negatively charged, as well as the Sun, to get the theoretically predicted shapes and effects. The "planets" Jupiter and Saturn were actually small star-like objects and had clearly measured "proton winds" (winds of positively charged protons) and, along with their moons, were exhibiting the same effects as the Sun and comets.

This led to the conclusion that the Sun's charging process was related to the solar wind and what followed had escaped every person who had ever studied the Sun and Plasma Physics. The Sun (and therefore Jupiter and Saturn) produced an excess current of protons in a solar wind that blew off their surfaces. This meant that the standard solar model that claimed the Sun's fusion was deep in its core was wrong, and that the nuclear burning was high up in the solar atmosphere. This immediately solved one of the long-standing dilemmas of the old solar model, the so-called "missing neutrino problem", since the fusion now took place at a slower rate. I knew I was on the right track.

It further explained that the Sun was internally acting like a "super atom", with the internal regions positively charged and a sea of high-energy electrons moving around it. This then solved other problems. One was the mechanism that held the solar corona in place (the observed sheath of high-energy electrons that surrounds the Sun). It also showed how the solar wind was formed as the fusion reactions in the solar atmosphere were ignited by intense electrical discharges in violent atmospheric storms, and hurled the composite (pinched) streamers of protons and electrons outwards. The electrons were retarded by the solar corona of electrons while the protons were accelerated as if through a high voltage screen.

When the proton and ion streamers emerged, they "opened up magnetic holes" in the corona and blasted outwards into the solar system. This left the Sun with a net negative charge and a sea of available free electrons as required by my model and the new data from spacecraft. This produced and maintained a "capacitor" ... and I called it the "solar capacitor". In the cases of Jupiter and Saturn I called it the "proton wind supported capacitor". This required further that there must be continual return current sheets from the outer edges of this capacitor back to the central fusion star to balance the "circuit". It did not take me long to deduce that this occurred in the direction from the Sun where Earth would be in the month of August, the month of maximum auroras and hurricane activity.

There also had to be a surrounding "doughnut nebula" of dust and gasses that had been swept out of the proton wind region and held back by the constant pressure of that wind. This became the first known source of comet tail material as the new comet theory developed. Later, many sources of comet tail material were identified including the rings of material and free ions that existed in the solar system that would combine (.e.g. hydrogen and oxygen to form water, hydrogen and carbon to form hydro-carbons, etc).

Remember that this was coming to me in 1980, 20 years before NASA's big push to understand "The Sun-Earth Connection". Magnetic fields that were then being discovered for the first time, were caused by ongoing electrical currents as the planets continually discharged the solar capacitor. I began to predict severe weather based on the electrical alignments of planets and especially comets. The fact was that these magnetic fields became compressed as "solar storms" hit, causing the transverse electric field to compress which in turn compressed our ionosphere which in turn drove the electrical currents that powered the severe cyclonic storms on Earth's Surface.

No wonder the weather men could be right one day, and totally wrong the next. Their weather models did not include the most significant factor in all of weather … the solar electric currents. I realized that the Sun was producing far more energy in the form of electrical energy than it did in the form of light energy. We typically were buffered from this by our protective magnetic field, but when solar conditions changed, Earth was subjected to enormous sources of power. It started to become clear, why use solar cells when there was such a simple and much larger source of energy?

I then realized where the vertical electric field came from that Tesla tapped into when he experimented in the early 1900's. It was indirectly powered by the solar capacitor. I had discovered the source of unlimited power that was available for free … we no longer needed oil or fossil fuels or nuclear fuel. There was a simple way that spacecraft could tap into this also. I attended some conferences devoted strictly to the topic of Electric Propulsion in Outer Space, only to find that these scientists were as cloistered and removed as the scientists I had met at NASA.

I began to develop methods of controlling and reducing the energy of damaging storms on Earth (primarily hurricanes and tornadoes), and developed methods for converting that power into useful electricity that could be distributed directly on our currently existing electrical grid system presently used across the country. A simple model showed that these storms formed when electrical currents connected between the ionosphere and the tops of clouds.

This is also where the lightning came from. I never did buy the idea that clouds could develop electrical charges internally that discharged ***outbound*** to other regions. I also calculated that the warm water theory for hurricane development lacked sufficient energy to account for the energy in these massive storms. We later witnessed hurricanes on Mars where there is no water at all. Clearly the warm water concept did not work (even though it is still in use today by NOAA and our national weather service).

I later learned that during severe solar weather, a good deal of water in storms actually is coming from outside Earth's atmosphere (from Earth's "comet tail") as the electrical connections occurred between outer space and Earth's upper atmosphere. It started to become clear that clouds of water in large storms could form where no atmospheric water sources existed (e.g. from ocean condensation). Much of this excess water ends up on the Earth's polar caps.

I further realized that Earth's jet streams were powered by the solar electrical currents, and that there were 3 belts of electrical current circling the globe. Two "ion" current sheets traveled from west to east at higher latitudes (in the northern and southern hemispheres), while an equatorial "electron" current sheet ran east to west. These "current directions" add a non-permanent magnetic component to the permanent magnetic field of Earth. Traditional theories state that the Sun heats the Earth's surface and this in turn drives the upper level jet streams ... this is actually backwards from reality. It is the equatorial electron current sheet of the equatorial jet stream that powers the hurricanes and typhoons that move westerly along the equatorial zone. It is actually the electrical currents, caused by the vertical electric field between the ionosphere and the cloud tops, that power the storms ... a complete model of how this works is beyond the scope of this book).

I then learned that the reason hurricanes lost power when they approached land was that the powering electrical current from the ionosphere to the cloud tops and to the Earth's surface had no connection (anode) while over the ocean ... so it drew up vast surface areas of ionized air from the ocean surface and sucked them up a central column (the spinning vortex was caused by the moist air rising "up the drain")... whereas the land provided a "ground" for the current and therefore it shunted out the storm's power source. As mentioned previously, I was now able to predict accurately, sometimes months in advance when there would be severe weather based on electrical alignments of planets, our Moon and comets.

In conjunction with some colleagues, in 1996 I offered to give this information to NASA and NOAA for free in an effort to stop the damage occurring around the world due to the increase in severe storms that have occurred since the mid-1990's. I have never heard a response from NASA or NOAA. They continue to take all the federal funding and use their incorrect theories as millions of people's lives world wide have been devastated by severe storms. Subsequently Hurricanes Mitch, Floyd and hosts of others around the world tore apart the lives of millions of people.

I then realized that the Great Red Spot on Jupiter (and the many smaller storms as well as the "sun spots" on the sun) are nothing more than ongoing hurricanes, but since there are not land masses to "stop them", they keep going on forever, and continue to circle in these giants in their huge stormy atmospheres. The giant gas "planets" and the Sun had many counter-rotating rings of electrical currents that are just like the three "jet streams" that circle Earth.

I realized that the high and low pressure cells in Earth's weather are direct results of current sheets connecting with the Earth's upper atmosphere and "streamers" in Earth's comet tail. I also learned that the jet stream electrical currents snake around the high and low pressure cells as they wind their ways eastward or westward. These reacted in kind to solar electrical storms and in turn drove the lower level weather systems.

It also became clear that solar storms affected many surface conditions on Earth. The most noticeable were the increase in earthquakes and volcanic activity. Another was a bizarre phenomenon called "El Nino". Once again scientists looked for "internal explanations" for any associations. They noticed that when El Nino was present, hurricane activity was affected. This would be like noticing that there is an increase in summer temperatures when the price of refrigerators goes up. There is a relationship, but it is not direct.

There is a relationship between these factors, but it is related to the source of power, the solar electric field. The topics of increased volcanic activity and earthquakes, is deferred to the later chapter on these specific topics. But El Nino weather is related and introduced here. The theory that El Nino is caused by a surface wind effect once again does not include what every well trained physicist is taught to do as a part of any analysis, that is to do an energy calculation. In the case of wind blowing warmer water around, it cannot account for a 2 to 3 degree rise of thousands of square miles of ocean in a few hours (the extreme case test is one of my three tests of a theory). Since it is surface heating only, most scientists will agree that it is not caused by any heat source from below such as volcanic venting. The only remaining source is the vertical electrical currents in the Pacific (which has no land masses) and is caused by what is called "joule heating".

More simply put, it is due to the heating caused by electrical currents from the ionosphere. It is then related to the hurricane activity in the Atlantic Ocean since the hurricanes also derive their power from the same source. The difference is that the Atlantic has Africa to the east over which the storm cells develop that catch the vertical electrical currents so that when they spin out over the Atlantic, they are ready to develop into hurricanes. The Pacific Ocean has no such counter part and therefore El Nino heating is the result.

Numerous references have been made to long term predictions of severe weather and hurricanes. It should be mentioned that my methods of predictions are based on the Plasma Discharge Comet Model and the new understanding of how Earth's electrical alignment with the Moon, other planets, comets and the electrical current sheet(s) that flow to the Sun from the solar capacitor. I will give two examples here of exact (do the day) predictions that were documented months before on my internet site.

This method of prediction is different than that used by the National Oceanic and Atmospheric Administration (NOAA … also called the National Weather Service), which simply takes statistics from previous years and extrapolates them into the future. Of their published list of "factors that cause and affect hurricanes and severe weather", none even remotely hint at solar weather (although they are scrambling as this book is published to include such factors) and certainly none refer to alignments with planets, the Moon and comets. Their method would be akin to saying that based on previous years of winnings of casinos in Las Vegas, this years winnings would be such and such. They do not predict dates of severe expected weather, but give an overall number of severe storms that will occur in an entire season.

We have seen repeatedly that they watch this carefully as the season progresses, and as in the 2001 season when there were too few storms, they started classifying any little breeze as a "tropical storm" so that at the end of the season their "prediction" came out correct. Their predictions are a cop out at best and serve no real purpose in protecting the public. I could write an entire book just on the fiasco that is the National Weather Service. But to serve this purpose here I will include one example where I predicted to the day the severe weather. One could see the solar flaring occurring just as I predicted as the comet – Earth – Moon – planet alignments came to pass. Below is the exact story as told on the internet in June 6, 1999 (the prediction) and September, 1999 (the predicted date) came to pass along with the absurd reaction of the National Weather Service. Note that the hurricane developed on September 6 as predicted, and as I predicted in June, there were no less than 4 major hurricanes developing in the Atlantic at one time. The story takes place a few days after September 6 as the first storm approaches the south Florida coast.

YOU CAN'T PREDICT THE WEATHER IF YOU DON'T KNOW WHAT CAUSES THE WEATHER
(or why are children in Florida in school as a CAT 5 hurricane approaches off shore?)

By James McCanney

The morning of September 13, 1999 ...

I recently sat in amazement at the image of the now famous Dr. Gray (chief hurricane scientist at NOAA's hurricane prediction center) as he was interviewed on TV with the ordained title "HURRICANE EXPERT" pasted across the bottom of the screen. This was an interview in response to the second time in as many years as the official predictors failed to warn residents of the mid Atlantic states of impending danger (in this case hurricane Dennis as it reversed directions and returned to shore).

This morning Monday September 13, 1999, I was horrified as I called a friend living in southern Dade County Florida, to find out that at 10:00 AM no evacuation warnings had been given and that children were in school. Hurricane Floyd, which had just turned CAT 5 with sustained winds of 160 mph and gusts nearing 200 mph and whose monster size far exceeded infamous Hurricane Andrew in all respects, yes Hurricane Floyd could be and probably would be encompassing all of Florida in the next 24 hours. It was not until nearly 11:00 AM that the Dade County mayor gave even a hint... yes Dade County could be in trouble.

The problem was that 1) there was no evacuation plan and 2) the only window of escape for all of south Florida was this day since as night would fall, the storm's initial rain, wind and high seas would preclude any future escape. With seas exceeding 35 feet on the northwest arm of Floyd, the only escapes would be the inner state roads. If the hurricane turns southward all of Miami and south through the keys would have little chance of survival. If an evacuation were ordered right now, the traffic jams would preclude most of southern Florida from escaping before nightfall!

The problem turns more severe when you see that there are 2 very large hurricanes on the heals of Floyd. One has just passed hurricane status and is only about 1/2 of the way across the Atlantic and is sufficiently south that it could follow in Floyd's footsteps (in fact according to "official" hurricane "science" this cannot happen ... more on that below). The point here is that no matter where Floyd hits, the rain and wind and tidal damage will preclude most survivors from fleeing any areas south of the damage prior to the next storm's arrival if this probable scenario unfolds. As my friend said, we might be able to survive one hurricane but a second one in succession would be lethal.

So why are children in school today on Monday September 13, 1999? Why has no evacuation plan been announced and why has no evacuation been announced up to now (just hours before dark)?

ABNT20 KNHC 131503
TROPICAL WEATHER OUTLOOK
NATIONAL WEATHER SERVICE MIAMI FL
1130 AM EDT MON SEP 13 1999

FOR THE NORTH ATLANTIC...CARIBBEAN SEA AND THE GULF OF MEXICO...

THE NATIONAL HURRICANE CENTER IS ISSUING ADVISORIES ON HURRICANE FLOYD...LOCATED ABOUT 525 MILES EAST SOUTHEAST OF MIAMI FLORIDA AND ON HURRICANE GERT LOCATED AROUND 1400 MILES EAST OF THE LEEWARD ISLANDS.

A TROPICAL WAVE WHICH HAS BEEN MONITORED FOR THE PAST SEVERAL DAYS OVER THE ATLANTIC HAS BECOME BETTER ORGANIZED. SATELLITE IMAGERY SUGGEST THAT A CIRCULATION MAY BE DEVELOPING ABOUT 800 MILES SOUTHEAST OF BERMUDA. SHIPS OBSERVATIONS INDICATE THAT STRONG GUSTY WINDS ARE ALREADY OCCURRING TO THE EAST OF POSSIBLE CIRCULATION. IF THIS TREND CONTINUES...A TROPICAL DEPRESSION OR A TROPICAL STORM MAY FORM TODAY OR TUESDAY.

The second part of this article is to recall the efforts my group made last May and before - at announcing that, following September 6, 1999, this would be a treacherous time for Hurricanes and we could likely see multiple large Hurricanes in the Atlantic that would rival Mitch. Hurricane Mitch of the 1998 season, devastated Central America killing tens of thousands and leaving millions homeless ... and whose remnants were felt as far north as Kansas and Minnesota in the weeks following its passage through Central America.

Here's my exact quote from my first report on Comet Lee and my predictions of what might occur:

Date: Mon, 07 Jun 1999 01:24:32 -0500

I've been able to pull quite a bit out even with the bad or unreliable data available from NASA and NOAA. Besides the mid 2000 alignments already mentioned, what really stands out is Sept. 6 1999. This will be a date when the Earth crosses through the cometary orbital plane of Comet Lee and Venus is in exact alignment. Almost spooky how exact it is! The comet's wanderings will be primarily within its current orbital plane and if what seems to be shaping up is true

regarding the reduction in orbit of the comet (falling inwards towards the Sun and slowing down), and given the circa August solar current sheet (solar tail) location, coupled with high CME potential, as well as the post comet perihelion solar increase in activity that is expected--- This entire time frame of August to September, 1999 could be a very serious cyclonic storm, and earthquake, era as well as general other effects such as the opening of the Earth's magnetic field to allow high energy particles to stream in to the planetary surface. Oddly enough the lunar alignment will also favor increased potential for Earth magnetic field break down at just that time. The Moon will be passing new-moon stage which will greatly increase the effects of the solar flares. It is almost as if someone or something had planned this! We really need access to good data and computer processing to determine the possible time frames, and truly locate this comet and its historical path to help predict the future.

We have to locate the episodes of wandering and try to predict the future. Too bad NASA has destroyed the Hale-Bopp data that we could really use to help provide the correction factors needed on Comet Lee. We also have to search for follow on comets as they usually come in multiples. We really need resources to go any farther and this is shaping up to be very serious. There is no time to waste, and if the past is any indicator of the future (i.e. Hale-Bopp) the NASA/JPL boys already have the good data buried so deep we will never see it. I wonder what the Hubble Space Telescope has been looking at lately? We do need resources."

Jim McCanney

The Aurora index of 10 was recorded on the very day that Floyd began its intensive growth cycle. An activity level of "10" is the highest currently possible level! In fact an activity level of 10 is quite difficult to locate historically!

I believe also that it is important that this was not guess work but was an exact calculated prediction based on the electrical alignment of Venus, our Moon (at new Moon which we have discovered is the source of heightened solar wind activity at Earth) and the Comet Lee which has maintained a VISIBLE electrical connection to our Sun since last May.

At the same time, a graph of the solar "k" index revealed the heightened solar activity present here at the Earth. As the Sun and the comet continue to pump out electrical energy and these levels stay high, we continue to be in danger of further natural disasters.

The point of this article is that the official national weathermen do not actually know what causes weather and are totally off base regarding the energy sources that drive hurricanes and typhoons. At this time of year the storms should be decreasing in numbers, size and strength, not increasing.

In addition to stating what the true cause of energy is that drives these storms, the end result is that it is now known (the knowledge has existed for years now) how to dampen the energy from these monster storms. We have had an open invitation and in fact have implored the "official weather centers" to use this knowledge for free to limit the damage of these storms. In their arrogance and self righteousness they have categorically refused.

You cannot predict the weather if you do not know what causes the weather. This is why the weather men seem to be "right" part of the time and so clueless on the following day. As you read the past installments of this page you will realize that as a result of the knowledge gained about how our solar system works with the electrical nature of comets (using my Plasma Discharge Comet Theory), we have also been able to predict 3 months in advance the exact timing to the storms you are now witnessing. Some might say chance, but look at the Solar and Magnetic readings. I can now say with confidence that there is no doubt that the current hurricane structure you see in the Atlantic Ocean is a direct result of very predictable factors given the new understanding of the Electrical Nature of our Solar System.

End of this historical account

One final note on hurricane Floyd. The National Weather Service, using computer models, predicted that Floyd would turn north and it did. But they are playing a dangerous game of Russian Roulette with the public. If another factor had changed only slightly and they were wrong, millions of people would have died in what would have been the worst disaster in American History. Numerous high ranking NOAA officials quit after Floyd, something that is not well known.

After the near miss of Florida, Floyd went out to sea. NOAA had the ongoing offer of my free information to dampen this storm and rid it of tremendous quantities of rain before it hit the Carolinas and eastern seaboard states all the way to Nova Scotia. NOAA again refused to accept this offer, and Floyd went on to destroy state after state with the most devastating floods in the history of the affected areas. Many people in those areas have never recovered.

A second account of unusual comet – Earth – Moon – planet alignments occurred in the summer of 2001 when a newly discovered comet … Comet C/Linear A2-B arrived in the solar system. I then had internet broadcasting capability on my web site jmccanneyscience.com and this prediction is documented there. This is but one more of the many predictions I have made using the electrical nature of the solar system.

This series of no less than 6 alignments in a few month period provided an amazing test that again worked out to the day. As each predicted day arrived, the solar flaring began and within a few days the solar flares were recorded at Earth and the storm activity began.

On the first date the National Weather Service made no prediction of severe weather. Suddenly and seemingly out of nowhere, massive mile wide tornadoes formed in Minnesota and ran for nearly 100 miles along the ground into Wisconsin, destroying entire towns in their path. The weathermen stood confused that their bright sunny forecast turned into the most severe storms ever witnesses in these communities that were caught totally unprepared. Even as the storms developed and raced across the states, no warnings were issued.

In July of 2001 I was studying comet C/Linear 2001 A2-B as its orbit brought it up and behind Earth just as the path of comet Lee noted above. Once again the alignments were uncanny as if someone had sent this comet in to teach man a lesson. The list of alignments (with the Sun in the center) is as follows:

- June 13, 2001 Quadruple alignment
  Mercury - Earth – Mars – Comet
- July 20, 2001 Triple alignment
  New Moon – Earth - Comet
- August 8, 2001 Quadruple alignment

Earth – Comet – Mars – Solar electrical current sheet

- August 18, 2001 Quadruple alignment
  New Moon – Earth – Comet – Solar electrical current sheet
- October 2, 2001 Triple alignment
  Mercury – Comet – Mars
- October 14, 2001 Double alignment
  Mercury – Earth - (comet is now too far out of alignment)

The June 13 alignment produced a visible flare and electrical connection between the planet Mercury and the Sun, and as already mentioned, the Sun produced an M7 class flare (the largest flare possible to date) that included a massive dose of x-rays. It took a little over 3 days for that flare to hit Earth and ignite the huge tornadoes noted above (as well as other storms around the globe).

Once I realized the basis for the true electrical nature of the solar system in the late 1970's, the domino effect of the Axioms of Astronomy started to topple. The standard solar model was incorrect, space was not electrically neutral, comets were not dirty snow balls but were a complex plasma discharge of the solar capacitor, comets were highly energetic objects and could become large enough to become planets and moons of the solar system (Immanuel Velikovsky may very well have been right about Venus being a huge comet ravaging Earth only a few thousand years ago), the solar system did not form all at one time 4.5 billion years ago, the standard explanations for severe weather were wrong. The list continued ... the explanations for the source of oil and fossil fuels were incorrect, the theory of slowly migrating plate tectonics and mountain building was incorrect, the "Greenhouse Effect" did not work on Venus so it certainly could not work in Earth's thin atmosphere. With respect to weather predictions, NOAA and the National Weather Service computer models were missing the most important factors that affected severe weather.

Many people had observed some of these details before, but what my theoretical work now provided was a firm end to end explanation for how all of this worked. HOWEVER, there was just one small problem. Every time I attempted to use the standard channels to release this information, it was immediately blocked. My papers to scientific journals would be returned without ever being sent to a referee. I did notice that after my papers went to certain journals, what appeared in my writings slowly began to turn up in the journals, but always somehow mixed into the standard axiomatic structure.

My access to the news media was blocked by NASA. Although I belonged to the international scientific organizations with full rights to attend regular conferences, I was never allowed a speaking spot. At most I was put in a corner in a "poster session" and the senior scientists made sure that the junior scientists would not talk to me. Official news conferences were scheduled behind closed doors.

This has placed the scientific community in a very bizarre position. They want desperately to explain the "Sun-Earth Connection", but their basic theoretical structure is incorrect. Furthermore, they have to deal somehow with the fact that I have already worked out all the details based on an entirely new set of basic principles, which are not at all compatible with their view of the solar system. They additionally have to deal with the fact that millions of people are receiving this information now on the Internet, on radio and TV programs and in this book. The public now has access to this information, and NASA cannot do anything about it, although they are trying numerous tactics daily.

The "Sun-Earth Connection" is very understandable in light of the Plasma Discharge Comet Model and its redefinition of the Sun and its solar capacitor. Now add a large new comet that has intentions of becoming a massive new planet or moon in the solar system and the term "Earth Changes" takes on a dramatic new meaning for people who have not been exposed to these concepts previously. The following brief chapter deals with the massive comet Hale-Bopp since it illustrates the numerous effects at a distance that affect our planet, and will contain some little known facts about the comet that graced our night skies for a number of years in the late 1990's.

IX. COMET HALE-BOPP AS AN EXAMPLE

Comet Hale-Bopp (hereafter referred to as H-B) is a prime example that showed us how a large comet could pump up the Sun on an ongoing basis for years and dramatically affect Earth's weather although it was hundreds of millions of miles from Earth. It was the primary cause of the extension of the solar cycle for at least 2 years beyond the year 2000 (something that had never occurred before). As it crossed the ecliptic (plane of the planetary orbits) it exploded into an electrical inferno.

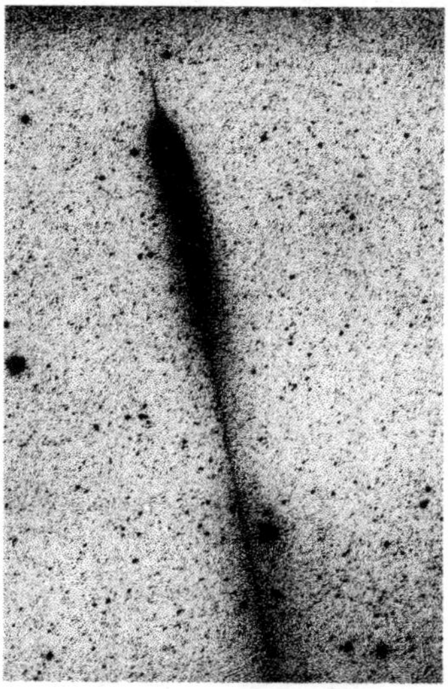

The Huge Comet Hale-Bopp as it crossed Earth's path exhibits its "sunward spike" (again called an optical illusion by European and NASA astronomers). Note how it appears as the two comets in the Indian rock paintings on this book's cover. This photo courtesy ESO.

The most amazing aspect of this great comet was the levels to which NASA and the National Security Agency went to hide the data and truth about H-B from the public, an effort that is ongoing today. Its effect on international politics extended even to the Vatican which quickly funded and built a major astronomical observatory and hired professional astronomers to man it (located in the state of Arizona). A Vatican memo to NASA read, "Is Hale-Bopp Wormwood?" This of course refers to the many biblical references to the giant planet that ravaged Earth and "caused the poisoning of waters" (the literal translation of the word wormwood).

The list of lies from NASA was nearly endless. Almost no sooner than photos from the Hubble Space Telescope had been placed on the official NASA internet site, than they were as quickly removed. NASA time lapse movies showed clearly a large companion or satellite orbiting the parent very massive and huge main nucleus. I called my contacts at Goddard Space Flight Center and the inside story was that H-B was HUGE. The ensuing story of the companion (which NASA officially admitted later WAS THERE at the Canary Island H-B conference) is so bizarre that it will take a separate book to describe it. NASA scientists accused individuals including some of their own professional scientists of "faking photos" and causing the death of 39 unsuspecting Heavens Gates Cult members, when in fact it is now proven that the NASA scientists at the University of Hawaii were the ones who doctored the photos.

NASA is still lying about the size of the nucleus, which is in fact about the size of our Moon and is the newest member of our solar system. My theory was dramatically proven correct in all of its detail with H-B, yet NASA continues to hide behind the Dirty Snowball Comet Model (for national security reasons). Isn't that convenient.

The truth is that NASA, the NSA, and other government agencies are prohibited by law from disclosing to the public anything that would cause a national panic. So too they will try to prevent dissemination of my theories about comets because it might cause a public to redirect its allegiance as a new and potentially dangerous comet comes into the solar system. While the government officials are using tax dollars to build safety caves for their "shadow government" in case of "major disaster", they are leaving the public out to dry with no forewarning or protection.

They know that H-B holds the key to understanding that their scientist's theories are wrong, and my theories have been 100% on target. But only a few core scientists in NASA are aware of this, and many of the lesser ranking scientists still believe that their theories are correct. This is due to tight manipulation and control of "sensitive data" even within these agencies. Yes, comet H-B holds some secrets.

The first is that it electrically pumped up the Sun for a period of about 6 years, causing near solar maximum solar weather during the solar minimum as it passed through the inner solar system in the late 1990's. The solar cycle did not end. It continued and the Sun in 2001 and 2002 hit record levels. A news release (March 21, 2002) showed NASA scientists stating that finally the maximum was reached and would turn down at that point. This was wishful thinking as the Sun continued to rise to new and unheard of levels of flaring and powerful storms. In April the Sun looked like a popcorn popper, blowing out flare after flare. Was it reacting to a new incoming comet (since H-B was now far on its way out of the solar system)?

As this book goes to print, all the major observatories of the world are being taken off line. Astronomers are being told not to discuss "Planet X" with the public. As with H-B, NASA has shut the door on release of information. They are positioning their scientists to become part of the nightly national weather programs, and to be in position to defray any public awareness of what is truly happening with the Sun and our planetary system.

The topic of H-B would take up a volume. My theoretical work was at the center of the huge ongoing controversy that raged on internet sites in the 1990's. But one single fact about H-B should make even the most ardent traditional astronomer stand and take heed. On January 2, 1997, H-B crossed Earth's orbital plane. NASA scientists made a huge internet ploy showing that Earth and H-B would not collide. As usual, they purposely downplayed the results of my theoretical work and the fact that the Sun was at record activity levels. They did a good job of "dumbing down the public".

No one that I know of said anything about comet H-B colliding with Earth or even passing close. They once again defrayed the "action at a distance" concept. What they failed to say is the fact that if Earth were just 3 months earlier in its orbit, we would have passed a mere 1 million miles from the massive comet nucleus (which they are still lying about) and we would have been engulfed for many days and possibly up to a few weeks in the comet coma (clouds of dust and gasses of the comet). Even if you believe the dirty snowball model, we would have had the Sun blotted out and intense influxes of water and pollutants during this time. We would have experienced a vast pollution event and the temperature of Earth would have reduced into a mini ice-age. H-B was a HUGE comet and would NOT compare in any way shape or form to the passage of Earth through the tail of the tiny Comet Halley in 1910.

Earth would have sat directly in the sunward electron bean of that huge comet. We would have also had the full suite of effects described in this book from a close encounter with a massive comet. My personal estimate, including the tidal waves in the oceans and land, earth quakes and the wrenching of many volcanoes, the huge hurricanes and storms, the electrical discharges to Earth, the exposure to solar radiation as our protective magnetic field was disrupted, etc., … my estimate is that Earth would have had a severe mass extinction and pollution event. The few survivors would have been driven back to the cave ages.

Why was the Vatican so intent on building a state of the art observatory and manning it with professional astronomers? Is it that they had nothing else to do? Or were they seriously worried about the biblical planet "wormwood"?

NASA does NOT want you to think about that. They additionally do not want you to know that there may be a new comet on the edges of the solar system and that it is affecting the Sun already. As my theory and observation states, these comets come in multiples, sometimes years apart. Was H-B the little brother of Nibiru? Will Nibiru have other orbiting massive objects that may split free and wander through the solar system as giant comets of their own? I could help answer these questions, but NASA has shut down the sources of data again, and I doubt if I will ever be allowed to see them.

H-B does hold the key to understanding, and NASA has the door to the data locked.

X. POLE SHIFTS, MASS EXTINCTIONS AND EVOLUTION, THE ORIGIN OF OIL AND COAL, METEOR SHOWERS, FLOODS, ELECTRICAL DISCHARGES BETWEEN PLANETS, POLLUTION EVENTS,EARTHQUAKES, VOLCANOES, LAND HURRICANES, CONTINENTAL FORMATION, MOUNTAIN-BUILDING, THE CLEANSING OF THE EARTH AND THE REGENERATION OF THE EARTH

The previous chapters leading up to this point were designed to lead the average person to understand the workings of the solar system based on the Plasma Discharge Comet Model. If you skipped those chapters to quickly read this, then you will be sorely disappointed because you will not really understand what you are reading, One of the biggest problems with the plight of traditionally trained astronomers, astrophysicists and space scientists (which filters down to the general public) is that simply they cannot shed the old ways of thinking and at best will try to mix their old concepts with a COMPLETELY NEW SET OF AXIOMS. Change does not come easily.

Those who have been brain washed with the monthly propaganda posted in Astronomy or Sky and Telescope magazines will equally have a difficult time dealing with the fact that they have swallowed hook line and sinker the NASA set of axioms, and additionally have not read between the lines that data that is being reported is completely contradictory to expectations of traditional astronomers.

There is an unfortunate sentiment that has evolved, and unfortunately was promulgated by the late Carl Sagan, that man with his intelligence determines the correctness of scientific theories. I constantly receive communications from amateurs stating that they see, for example, comet tails separating from the main comet and being blown outwards, therefore "proving the Dirty Snowball Model". In fact both theories state the same thing regarding the separation of tails from a comet … the solar wind will in fact blow it away in both cases. Underlying their intentions is a hope that they can "prove me incorrect". In their haste and ignorance, and in the same mode as the professional astronomer, they are hoping to dismiss me and send me packing. This topic is far more complicated than that and, furthermore, the extremely detailed and meticulous experiments that will ultimately prove my theories correct are well defined and have never been performed by the people who hold the keys to the equipment that can do these subtle and unique experiments. Only by mistake, as with the discovery of intense x-rays to the sunward side of comet nuclei, have my predictions been proven correct.

In spite of this denial, sufficient evidence has been recorded to prove that the Dirty Snowball Comet Model is incorrect, and the Plasma Discharge Comet Model along with the new solar model is correct. Two scientists came close to making the "switch" and were about to go public, and many are firmly convinced that they were killed. One was Dr. Gene Shoemacher (after whom comet Shoemacher-Levy 9 was named). The other was Dr. Richard Herrington. Both had the potential to pull many scientists with them from the NASA ranks.

With this stated, I will briefly introduce the concept of the primary Earth Changes that were chosen for this chapter. Note that there are hundreds of more subtle effects and secondary effects too numerous to discuss here. Just to give an example, when the electrical connection between a huge passing comet and Earth would occur, the ground would start to erupt with an outpouring of insects that could not withstand the electrical currents flowing through the ground. Plagues of insects, frogs and other Earth inhabiting creatures would engulf the Earth. The Bible and ancient texts talk of such trauma, but modern science tells us that this is only folly and fable. The list of topics included here are the major effects, and hopefully another book will be able to go into more detail.

POLE SHIFTS – Of all the topics relative to Earth Changes, there has been the most confusion regarding pole sifts. There are two types of pole shifts. The first is magnetic and the second is a physical shift of the Earth spin axis. Without trying to explain the misinformation relative to this topic, there is a misunderstanding even amongst professional astronomers about the true nature of Earth's magnetic field. I have devised a 5 layer magnetic field model for Earth. Some planets are less complex (e.g. the planets without moons).

Earth has a small central core (the original comet nucleus when Earth formed as a planet from a huge comet probably billions of years ago). This core has an inherent permanent magnetic field and from its density one must presume that it is composed of iron and heavy elements. This is only a minor component of the total magnetic field. The layers of magnetic field moving out from this core are electrical currents in the molten sub-mantle, deposits of iron and nickel in the crust and mantle, electrical currents in the ionosphere (the jet streams), electrical currents in the Earth "comet tail" that attach to Earth's upper atmosphere and cause high and low pressure cells, the Van Allen Belts, the electrical currents in the overall flow of electrons coming from the Sun and ions coming from the non-sun side of Earth as well as negative ions moving away from Earth (yes Earth is a comet), the outer shell of Earth's overall magnetic field that interacts with the solar wind, and last but not least the solar wind currents of protons that are building up the solar capacitor.

Most of Earth's magnetic field is of the non-permanent type, that is, in the form of electrical currents that are ultimately powered by the currents passing by Earth in the solar wind. If the Sun turned off today, the Earth would turn quickly into a frozen ball of ice with a minimal magnetic field. So when people claim that Earth is expected to have a magnetic reversal, this is not that unusual and it depends more on external conditions than on any properties of the Earth itself.

In fact in the late 1990's a huge solar flare completely nullified Earth's magnetic field for a number of hours, allowing the radiation from the solar wind to directly impact Earth. Russia at the time was facing the Sun. Russian scientists are tracking the cancer rates to determine the long term effects.

The magnetic field that one measures on the surface of the Earth with a compass is primarily due to permanent magnetic field components that are locked in the Earth's core and mantle in the form of localized iron and nickel deposits. The "north pole" of this component of the magnetic field is near Baffin Island more than 15 degrees from the "true" or rotational north pole. The "true" north pole is where the axis of spin of the planet resides (if you put pins in the Earth, the Earth would rotate around these pins).

There are many local deposits of iron and nickel in the Earth and these are NOT connected as in a large permanent magnet. The Earth's structure is not one of a solid rock. If you cut a section of the Earth's mantle with a giant saw and tried to pick it up with a giant crane, it would crumble and fall apart in millions of pieces. So these localized pieces of the field would not "flip" as a whole. What will occur (as discussed below) is that the magnetic "jolt" that these localized regions would receive could trigger earthquakes and volcanic activity as we have already witnessed when large solar storms hit Earth.

The "magnetic flip" scenario is one of the most commonly heard misconceptions regarding magnetic fields and the "pole shift". Additionally the Earth is buffered in outer space by a non-permanent magnetic field that constantly adjusts and changes to external conditions. The permanent magnetic field formed during a time when the Earth was molten and cooling, and there was a sustained EXTERNAL magnetic field (caused by the solar wind conditions) that remained there as the various regions of the Earth cooled through what is known as the Curie Temperature (this is how we make permanent magnets in a factory or laboratory).

As long as we are on this topic, there are a number of other pervasive incorrect concepts about gravitational fields. Some say that gravity is a secondary effect of electrical fields. This could not possibly be true since all the celestial bodies are surrounded by what are known as Debye shielding sheaths (they are electrically shielded by plasma) which buffer and constantly change. If these "electrical gravity" concepts were true, then satellites would not see our gravitational field due to the Debye sheaths and would go spinning off and never return or at best would orbit erratically and we would never be able to predict their orbits within even a few days. There are small very subtle effects that work over long periods of time (some long range satellites are not exactly where they are supposed to be) but these are extremely minor effects (not within the scope of this book). People can talk about such concepts all they want, but no one has ever shown experimentally (where independent scientists reproduced an experiment) that shows any relationship between Earth charging and the force of gravity. One does not need any of these types of concepts to understand the basics of "Earth Changes".

So there will be magnetic effects that are caused by a passing new planet or comet. They may be permanent or temporary, and the secondary effects (e.g. increased earthquakes and volcanic activity) may be far worse in the long run than the magnetic reversal itself. The entire topic as well as the history of magnetic reversal is very complex, but this will hopefully explain the basics.

Regarding pole shifts where the ***rotational axis*** of the Earth is affected. The last Ice-Age was accompanied by a radical pole shift. I always hear of traditional scientists coming up with bizarre scenarios for ways that the Earth could "flip" poles. As with most astronomers, few geologists have a serious background in classical physics.

The Earth will not just "flip" from internal causes. It is not a matter of finding enough energy to flip the Earth and then stop it after a 180 degree swap of north and south poles (which would in itself be astronomically enormous). The physics of such a flip would require something akin to a gymnast running and doing a flip and then stopping rotation in mid air, but with no floor to run on or stop on. The only reason these bizarre explanations are allowed in the standard journals is that it sets well with those who believe that the solar system formed 4.5 billion years ago and there have been no planetary intruders since. Yet they have to account for factual data that shows that Earth has had magnetic field alterations. To clarify this, there is no evidence in anyone's camp that says that the "true" north and south poles have flipped a full 180 degrees. There is evidence that localized areas (the so called magnetic moraines) show altered magnetic fields. The Earth has had lesser ***rotational pole shifts*** as discussed below. Basically people have been combining magnetic and rotational "flips" incorrectly. There is a relationship that must be understood in context. Expecting the Earth to "flip" either the magnetic field or the rotational pole by itself would be like asking a baby to carry a refrigerator up a mountain side. The scales of force are that remote.

Localized magnetic flips must occur when small regions (such as under an aboriginal camp fire pit in Australia) melted iron or nickel and it cooled in a short period of time when the external magnetic field of Earth was reversed during the cooling period. The other possibility for these types of areas is that during the wrenching of the Earth's crust during the passage of a large celestial intruder, the localized area was physically rotated as the Earth's mantle moved in the tidal wave that raced through the Earth's mantle.

The ONLY WAY that Earth can physically have its rotational or "spin" axis poles moved is by the close passage of a large celestial body. The last "ice-age" was accompanied by such a shift. The old "true" north pole was just north of the state of Wisconsin and Russian Siberia was a tropical climate with rain forests and mastodons roaming the forests. In a single night of total destruction, these areas were blown apart by incredible winds and storms and the pole shifted an estimated 30 to 40 degrees. Overnight the mastodon herds were quick frozen and buried in what is now the Siberian tundra with the fresh tropical plants still in their throats. This meat was so well preserved that it was used as food (and even sold on the commercial markets) in the last century during the building of the trans-Siberian railway. Even a single day of exposure to a warm temperature would have rendered this meat useless and rancid. The Earth Change was so rapid and complete, that this did not happen. The meat was as fresh as the day it froze, thousands of years earlier.

About a year ago a group of scientists pulled the remains of one of the carcasses from the tundra and are studying it. No word has ever been made public on their "findings", although it is speculated that they could be attempting to recreate the mastodon species and other species of insects and plants that went extinct suddenly at that time.

I have given a number of lectures on just the topic of the last rotational pole shift and the demise of the continent of Atlantis, which coincided with the demise of the mammoths. The cause of pole shifts was also defined in my original papers published in the early 1980's that are reprinted in Appendix 2 at the end of this book

Essentially the Earth's structure is as follows. There is a dense central core (the original cometary "seed" nucleus that has been measured by soundings). This also shows how much of a planet forms on top of the original comet nucleus. The Earth formed as a huge comet and is a very old planet (as opposed to Venus which is new and still molten with its huge volatile atmosphere, just out of the cometary stage of development).

Around this core are densely packed layers of heavy elements and composite materials that were sucked in during the comet stage of

Earth. Outside of this is a molten layer that is under the Earth's relatively thin mantle. It is the outer two layers that are important to the "flipping". The inner core regions are relatively solid due to the tremendous pressure. They will continue to rotate in the same direction and constitute the majority of the "angular momentum" of the Earth. When a new large intruding celestial object passes near by Earth, tidal waves will be set up in both the oceans and the mantle which, as mentioned above, is loosely held together (this is why the concept of plate tectonics fails ... the plates are not solid and rigid over the expanse of thousand of miles). The waves ripple through the mantle at thousands of miles per hour as the remote object's gravitational field drags the tidal bulge around the globe. It will slip over the viscous layer and when it stops it will resume rotating in the same direction as the inner core which continued to rotate in the original direction. The number of days per year will change and calendars will be rendered useless. How far the pole will move and in which direction will depend on the path of the external object. The Earth is not going through global warming or seeing plate tectonics. In both cases the Earth is "relaxing" from the destructive event of just a few thousand years ago.

When I have been on open mic radio shows, one of the most frequently asked questions is regarding where the poles will shift. There is no way to predict. There are so many factors. Even if we knew exactly the orbit of an incoming object today, its orbit would be changing on an hourly basis in an unpredictable manner due to the comet tail drag. It is clear that the last passage of a massive object which caused the pole shift (that moved Siberia into the arctic circle) came from over the north pole or under the south pole.

Tidal forces are short range "differential" forces caused by the difference of gravitational forces of the passing object on the near and far sides of the Earth. Only a very large object moving rapidly in close proximity to Earth would have any effect at all. Another greatly publicized misconception is that gravitational alignments of the planets cause earthquakes (electrical alignments do as discussed previously when large solar storms hit Earth as a result). Even the massive Sun's tidal forces are miniscule compared to those of our Moon due to the Sun's greater distance. Tidal forces are very short range forces. Only a very large object passing very close to Earth would have any effect at all.

This is also why the last ice-age pole shift is living proof that we had an encounter with a large celestial body at that time. This book will not speculate which object (Venus as a huge comet, Nibiru, Hale-Bopp on a previous visit, or otherwise, was the cause). There is evidence that the previous "true" north remained at the spot north of Wisconsin for a very long time. Indicators are the biological evolution of large animals in Siberia and the complete lack of animal life prior to that event under what we now call the Laurentian Ice Sheet. Further evidence is the condition of Antarctica which was in temperate climates prior to the shift and whose ice cap sets on a warm continental shelf (that is why it is sliding off and why there is a vast open unfrozen fresh water lake 2 miles under the ice cap called Lake Vostok). Direct measurements of the age of the Antarctic Ice Cap put it at no more than 4000 years old. Additionally, the closely coupled synchronized rotation of the planet Venus with Earth proves their recent close encounter had strong tidal forces. Traditional astronomers have always swept this contradiction under their theoretical rug.

MASS EXTINCTIONS AND EVOLUTION – The remaining topics have all been discussed by authors such as Donnelly, Velikovsky, and others. What this book has done is to explain why they occur when a large new celestial object comes into the solar system. Remember that ANY object large or small, coming in on an elliptical orbit, will discharge the solar capacitor and form into a comet. There are literally hundreds of factors that will affect the evolution of this object as it picks up material. The comet is a giant cosmic vacuum cleaner. All comets are driven by the electrical discharge of the solar capacitor and therefore will affect the Sun also (some more than others). The large ones will greatly affect the Sun even as they enter the far reaches of the solar system.

In spite of Carl Sagan's "calculation" in his "defeat" of Velikovsky at the 1974 inquisition sponsored by the AAAS (American Association for the Advancement of Science), Sagan stated that there was an infinitesimal probability that an asteroid or comet would hit Earth. The doors were no more than closed when nearly every scientist that was at those meetings embraced the "newly discovered fact" that the demise of the dinosaurs was caused by a catastrophe from outside.

NASA says that comets are tiny little "Dirty Snowballs" and that only if we are directly hit by a comet or asteroid will there be any effects. All of their concentration is on this. They are completely wrong on all three counts.

Although the majority of comets are small, some can be of planet size and it appears that they are not all that rare.

"COMETS ARE NOT DIRTY SNOWBALLS"
comets are a complex plasma (electrical) discharge of the solar capacitor caused by a rocky or asteroidal type body entering the region near our Sun – some of the objects may be well on their way to developing into planets from previous visits to the solar system

"Action at a distance" is by far more common and will drastically affect Earth and its inhabitants, sometimes when the comet is nowhere near the Earth by causing solar flaring, electrical discharges and electrical alignments between the planets, etc.

THE ORIGIN OF OIL AND COAL – Many authors have made it clear that by simply observing coal one sees that it is made up of leafy material piled up in massive amounts and the leafs are in a sense fossilized in these massive deposits. The material did not drop from trees, decay and then turn into coal. The first words in Ignatious Donnelly's book on the topic of coal indicates additionally that it was deposited repeatedly in layers with rock and other debris layered between the layers of coal. Coal was not deposited in a slow process, but was piled high and deep in a single process that did not take more than a few hours and took out vast areas of the world at the same time.

Oil is likewise found the world over in vast reservoirs under the ocean floor. Where were the fern forests that gave rise to these oil deposits? The answer has been clearly stated before. What this book does is state the observation that comets are not dirty snow balls and that in fact when Earth was in its own comet stage vast amounts of hydrocarbons were accumulated and are surely locked deep inside. The majority of the surface and mantle oil we are drilling for today came in as vast pollution events when LARGE comets passed near and Earth became a secondary "comet" in the discharge of the solar capacitor. Earth experienced vast pollution events in short amounts of time. This is also why oil is found to be thick in some coal beds ... because they were accumulated and formed at exactly the same time.

METEOR SHOWERS – As explained in the papers in the Appendices, meteor streams are associated with comets because of a subtle and minor force in nature called the "induced dipole force". It acts on objects containing metals and is a short range force. This is why meteors associated with comets are found to generally have metals in them. So a comet is not only drawing in large amounts of fine dust and molecules, it is forming streamers of "followers" in the form of meteor streams. Large comets with larger electric fields will be able to draw in huge boulders by this method. The ancient tales of fire and brimstone raining in during the time of the Great Flood saw vast amounts of rocky matter raining in and igniting the "naphtha" (oil and hydrocarbons) as they smashed to Earth's surface.

FLOODS – There are numerous types of flooding associated with comet encounters in addition to the oil and chemical floods from the encounter with the dense comet tail of a LARGE comet. Recall that the small comets visited by NASA to date such as Halleys, Borrelly and Giacobini-Zinner were extremely small and in no way shape or form could be compared to a massive giant comet such as Hale-Bopp, Venus (when it was in its comet stage) or the estimated size and power of the Egyptian story of Nibiru.

The types of flooding include the vast amount of new water that would come from outside the planet during the comet encounter. Biblical records indicate a rain of "40 days and 40 nights", which is about the estimated time it would take to pass through a comet tail that would be moving with Earth's orbit and an estimated 20 to 40 million miles across. Remember also that Earth will itself become a "comet" during this time so it will continue to drag in pollution even after the comet encounter has passed.

The lack of Sun for this length of time will additionally cause a general cooling and possible mini-ice age. The overall effects will be devastating and would render any transportation methods inoperable.

The second types of floods would be from local storms caused by the massive hurricanes and cyclonic storms that would form and cover the Earth. In the alignments with recent small comets and the planets we saw as many as 5 hurricanes in the Atlantic at one time along with multiple typhoons in the Pacific and Indian Oceans. As will be seen, Land Hurricanes will form also as low pressure cells sweep across the land masses of Earth. I am sometimes asked from what direction the winds will come or hear statements that a certain area would be immune to winds. The winds would come from all directions and there is no place on Earth that would be immune from their approach.

Coastal areas would receive enormous tidal waves if the massive comet/planet approached in proximity to Earth. The topic of land tidal waves will be covered in upcoming pages. As the areas under the ocean in future months and years crumbled and slid under the ocean, huge after shock tidal waves would continue to form for many years or even centuries to come.

A serious flooding condition, that occurred after the last comet encounter affected Earth when the rotational poles were shifted up to 40 degrees, was the melting of the old Ice Caps. The old north ice cap was what we now call the Laurentian Ice Sheet. As it rapidly melted, it carved out the upper Mississippi river valley whose span in some areas is10 or more miles across. To imagine that this was eroded by the small meandering river we see today could not possibly be true. The old South Pole was then in the south Indian Ocean region and as this moved into temperate climate the melting caused the ocean levels to rise quickly over the following decades.

The "fable" that the great continent of Atlantis (which this author is convinced was in the area we now call the South China Sea) sank under the ocean in a single night of calamity and destruction. This could be viewed as a form of flooding, and was probably a combination of tidal wave, massive Earth quake and raising of the oceans due to a massive influx of cosmic water. It should be noted that the US intelligence agencies have also been convinced of this location of the MAIN CONTINENT OF ATLANTIS (remember that Atlantis had colonies all over the Earth at its height of power).

ELECTRICAL DISCHARGES BETWEEN THE PLANETS - When Moses looked up as they approached the Red Sea, he recorded for history the "burning pillar of fire that extended into the sky". This was the cyclone that was the electrical discharge between the passing comet and Earth. My comet paper printed in 1980 referenced the first recorded example of the visible after effects of an electrical discharge that marred a planetary surface. Since, other authors have taken up the cause of researching and documenting other examples of such markings on planets, moons and asteroids.

The ancients also observed in many parts of the world the snake-like extension coming from a huge comet and "attacking the planet Mars". In a single stroke, this plasma discharge sucked the atmosphere and oceans off of our smaller brother planet. The "mouth of the snake around the planet" appeared as the intense electrical connection caused auroras in the upper atmosphere of Mars. My personal belief stands with that of Velikovsky in that it was the comet that became the planet Venus. My basis is strictly scientific. The chemical compositions of Venus and Mars are almost identical (although Mars of course has a much smaller thinner atmosphere). That is, the relative percentage of compounds found in the current atmospheres of Venus and Mars are almost identical, even though they have the very different planet Earth in between and exist in different areas of the inner solar system. When the pre-Venus comet encountered Earth, as already explained, Earth's surface gravity was larger than that of Venus and so Earth experienced a vast pollution event, whereas Mars, being smaller that Venus, lost its atmosphere to the larger comet nucleus that became the planet Venus.

POLLUTION EVENTS,EARTHQUAKES, VOLCANOES, LAND HURRICANES, CONTINENTAL FORMATION, MOUNTAIN-BUILDING - The close passage of a massive new comet as it forms a new solar system member will cause vast pollution events as Earth encounters the huge chemical filled comet tail. This is one reason why NASA is lying about the true size of the enormous comet Hale-Bopp. If people realized the vast size of that comet and the vast amount of matter in its tail, and that we came only a few short months from being immersed in that tail for a number of days, it would have been clear that their docile little snow ball model was wrong again. This would not have been like the 1910 passage of little comet Halley when Earth passed through Halley's small thin tail.

The magnetic jarring of the Earth's mantle, the tidal waves set up in the Earth's mantle and the electrical currents that would touch and ignite volcanoes into new eruptions are all effects that would occur on close approach of a new massive comet. As mentioned throughout this text, the comet might even be on the far side of the Sun and if it ignited the Solar surface to flare in our direction, the jolt that would pass through our multi-layered magnetic field could trigger earthquakes and volcanoes. The earthquake and volcanic zones of Earth are very much connected. Additionally they are like loaded guns. All it would take is a small force to "pull the trigger" and a chain reaction could ignite earthquakes and volcanoes that would in turn ignite more. The fragile Earth ecosystems, with this level of damage, would affect the lives of millions of people.

THE CLEANSING OF THE EARTH AND THE REGENERATION OF THE EARTH - There is scant enough room to detail all of the primary and secondary Earth Changes in this single 200 page book. The primary purpose of this book is to give the reader the scientific theory that is correct regarding the true nature of comets and their development into new planets, moons and smaller members of the solar system. In doing so the reader now can understand and put into perspective the other effects. Possibly a second book can follow to go into more detail on just the topic of Earth Changes.

One of the favorite lectures I gave in my college and university astronomy classes was the following. Take a look around you and see what exists today. Then go back 10 years and take away anything that changed in those 10 years, that is, view your surroundings as of 10 years ago. Then go back 20 years and do the same. At each stage I interjected comments on what we would or would not see. Then in a series of mental exercises I took them back 50, 100, 150, 200, 300, 500, 750, 1000 years and continued back to what we call "pre-history". The point was to simply view the physical world in these time periods.

We talked about the trees, forests and animals ... and most of all the civilization centers of the world. We estimated the populations and talked about the known separation of these civilizations and when in history they "discovered" each other. Many points become to emerge and solidify. The most striking is that in the timeframe of history, there was not enough time for the varying races (which already existed in their varying forms) to have evolved from a single central location as they had been taught as school children.

At this point they were figuring it out on their own. The history of the human race that they had been taught had some major flaws. It could not have possibly happened the way they were taught. They arrived at these conclusions themselves ... classroom after classroom after classroom. It was always interesting to watch the looks on these students faces as they saw everything they thought they knew crumble before them. The immediate response from the top to the bottom students in each class was the same. What were they being taught all this time and why could not the learned book writers and teachers see this most obvious of situations.

My lectures also sometimes delved into Christ's Sermon on the Mount or the Jewish tales recounted in The Legends of the Jews, especially the part recounted during the first Passover and the Exodus from bondage in Egypt. Many were amazed that these were passed off as fables and myths. But these varying stories told of the same events. They were not good times for planet Earth or its inhabitants. But in retrospect and in all religions or historic accounts, there is something good that came out of these most troubled of times as the Earth was cleansed and a new era began a fresh regeneration of our beautiful planet as well as a fresh start for its inhabitants

XI. THE HISTORICAL RECORD AS WITNESS

This book will not pretend to be a substitute or reconstruction of the tremendous work performed by other authors regarding the historical record. Every branch of study today includes catastrophes from outer space to account for those pesky anomalies that are most obviously only accountable by forces and intrusions from beyond the frail atmosphere of our water planet.

As pointed out repeatedly in this book, however, they all follow the NASA lead that it must have been a colliding asteroid or small comet nucleus that crashed directly into the planet. Even electromagnetic effects such as magnetic reversals are suggested to be caused by such collisions. I have yet to have anyone ***explain in terms of physics*** how an asteroid colliding into a planet causes a magnetic field reversal. But when no one referees your articles, as in the case with NASA publications in the journals SCIENCE and NATURE, you can say pretty much anything you would like.

The truth is that very few scientists within academic circles have looked at the overall picture that requires many simultaneous events of very diverse nature. How scientists can show a map of the Laurentian Ice Cap lapping far into North America, only reaching to approximately the current location of the North Pole, and at the same time showing Russian Siberia as a tropical climate ... and not come to the conclusion that within the past many thousands of years the Earth has experienced a rotational pole shift, is beyond me.

One thing is clear ... whatever the event, it 1) could only have been caused by forces from outside our planet and 2) it was NOT caused by a colliding asteroid (since there is no massive new crater). The process of elimination on just this basic level is clear. Earth has been subjected to a close encounter by AT LEAST ONE massive new comet becoming a planet in the time frame of no more than 10,000 years ago. The time frame that clearly makes sense for the last event is approximately 3500 years ago, the time we see a dramatic transition between "pre-history" and the digging of "modern man" out of the devastation. There is no longer any other possible scenario and modern science is impeding progress each day that it holds the reins of power and prevents this reality from blossoming forth.

XII. SPECIAL TRIBUTE TO IMMANUEL VELIKOVSKY

As noted in the last sentence of the last chapter, there is no doubt that Earth experienced the close approach of a massive comet nucleus as it evolved into a planet at least once in the past 10,000 years, with the most recent event most probably about 3500 years ago.

Rather that lash out any more at NASA (for wasting 30 years of the space program and a half a trillion dollars retarding true progress in science), I will explain exactly why Velikovsky was correct. His only mistake was in trying to explain his historical and physical findings using the incorrect theories that were provided by traditional astronomers. The other mistake of Velikovsky was in stating that the giant pre-planet comet "Venus" must have "popped out of Jupiter".

We clearly know today that the comet was ***CAPTURED*** by Jupiter, a process that is very common and well understood mathematically and has been observed, as every one of the major planets (Jupiter, Saturn, Uranus and Neptune) all have associated families of comets that were captured. The ancients saw this event as Venus the comet exploded into brightness as it left Jupiter. They watched as the new massive comet dominated the day and night sky for possibly many decades before it eventually passed Mars (previously a blue water planet like Earth) and lashed out with a huge electrical plasma connection to suck away the oceans and atmosphere.

Ultimately it is said to have come out of the direction of the Sun and was not seen by Earthlings until it was upon them. Within a 57 year period it passed Earth once again, and then, with the circularizing effect of the comet "tail drag", the waning comet settled into the orbit we see for Venus today. Venus is a volcanic hot new planet.

This scenario makes sense and is fully explained using the Plasma Discharge Comet Model. What is amazing more than anything is that all of this information was fully available over 20 years ago. What is even more amazing is that there is clear evidence that NASA is now hiding data that would prove that there is another massive object inbound into the solar system with potential for devastating events for planet Earth. There is clear evidence that they fully intend to set the world's population up to be BLINDSIDED by this object. Inside information indicates that NASA has "calculated" what they perceive to be the possible effects of such a passage. The problem is that they are using totally incorrect theories of how the solar system functions.

I could copy here the most recent issues of the American Geophysical Union newsletter that lists the ongoing conferences around the globe. Regarding the search for theories to explain the "Sun-Earth Connection", it is clear they are grabbing at straws and collectively have no clue what the underlying causes of space weather must be. As noted in this book, on a very fundamental level that is explainable and understandable by anyone with even a rudimentary level of scientific background, the fundamental theories of traditional astronomy are incorrect.

Velikovsky spearheaded a scientific revolution that should have followed his very innovative studies of the alignments of calendars of the many ancient cultures. The papers reprinted at the end of this book (the 3 part comet paper) show calculations of how a massive pre-planet comet such as the one that turned into the planet Venus would move and change its orbit as the orbit moved to a planetary type circular orbit. The massive comet Hale-Bopp proved this concept as its orbital period changed from 4200 years to 2650 years ***within a single passage.*** Welcome Hale-Bopp, newest planet of our solar system.

This is the dramatic effect that tail drag can have on a massive new comet. Hale-Bopp's orbit was changing so rapidly that NASA computers *could not keep up with it on a daily basis*. This observation alone proves Velikovsky's scenario of Venus nestling into a planetary orbit in less that 100 years, as every ancient culture reworked their calendars and religions based on the planet Venus.

Pole shifts, frozen mastodons with undigested tropical plants in their throats, electrical discharges spanning the heavens between the planets, ancients struggling to adjust old calendars ... and today we have teams of scientists studying the open fresh water lake (Lake Vostok) setting 2 miles under the Antarctic ice cap.

NASA scientists have been repeating in a sing song manner for 25 years that "Velikovsly has been proven wrong". It is time to set the record straight. It was NASA that was wrong. Velikovsky single handedly did more for the advancement of true science than all of the NASA scientists of the last 3 decades combined.

XIII. SUMMARY – WILL MAN SURVIVE?

The rules of nature are quite simple, and in the case of survival as an intelligent species (one that is capable of altering and controlling its environment) the rule of ultimate survival is clear and very simple.

As a species … collectively … if you do not use the time and resources allotted to you to gain access to outer space and gain the knowledge to survive … you will be doomed to extinction.

This clearly leads to some secondary basic rules of nature regarding the use of time and resources.

Petty wars and the use of resources to fight these wars can have only one possible result … and nature in its own time and manner will repay those species … collectively … who fail to control their leaders who promote these wars.

If there is a smaller group that attempts to use the resources of the many to save their own few and selective hides … these especially will be doomed to die a most lonesome and hideous demise.

The successful species that eventually will populate the stars will not allow those "unsuccessful" species to enter their hard-earned space and bring their petty wars and selfish ways into the cosmos. This is called the "rule of survival of the successful" … and is universally understood and commissioned to be executed upon those who fail at the first three rules.

These rules are stated to be inherent and self-evident. As a species we have failed. There are no less that 54 wars ongoing at the time of this writing. World leaders are out of control. As an observer of the Sun and comets, and a person with inside contacts to very high levels of the intelligence agencies, I will say that the writing has been on the wall for some time and there is little doubt that the leaders in these intelligence communities have been warned. Let the reader decide … based on the inherent rules of survival … will man survive?

XIV. APPENDIX 1 – OTHER SELECTED PUBLICATIONS

There is a tremendous significance in the following two short publications and I must say that I did not realize all of the implications when these were written over 2 decades ago. The "Galaxy" paper sets the stage for the Plasma Discharge Comet Model in that it shows how primitive solar systems form with two or three members ***at most*** in the beginning and add new members one by one at later dates. It showed theoretically why a solar system could not possibly form all at one time. This paper also resolved one of the long time mysteries of galaxies (the so called "missing mass problem").

It also presented a new form of orbital equation which removed time from the system and calculated in analytical form the orbits of all stars in the galactic arms for all time, a mathematical innovation that is still not understood or appreciated by the general astrophysics community. I also did not realize at the time the significance, but the Editor in Chief of the journal Astrophysics and Space Science (Zdenek Kopal) refereed the paper himself and assured its rapid publication exactly as I submitted it. This rarely happens in such journals.

The brief but very important "Saturn Sweeper Moons Predicted" paper holds what I still feel is the key to convincing many that the entire traditional solar system and solar fusion model is incorrect. The most important data that I feel was returned from the Voyager space craft was the fact that Jupiter and Saturn both exhibit proton winds coming off of their upper atmospheres. This should have told astronomers a host of items, the most important of which would be that these "planets" are undergoing nuclear fusion (something that cannot be accounted for in the standard solar model). Additionally, the resulting proton wind creates comets (the "spokes") in Saturn's proton wind supported capacitor which is like the solar capacitor of the Sun. This is HUGE!!!!

I believed then as well as today that the workings of Saturn prove that the basic traditional assumptions about the solar system are all completely wrong. This was dramatically confirmed when the 7 pieces of comet Shoemacher-Levy 9 approached Jupiter and released more energy than all the nuclear arsenals of Earth in each of the seven huge explosions.

One can see in the earlier chapter comparing traditional theory with the Plasma Discharge Comet Model, that there is no middle ground. Although the two sets of concepts claim to explain the same phenomenon, there is no way to merge the two because they claim diametrically opposing basic rules.

This is unusual in the history of science. Typically new theories add a level of complexity and elegance onto an already existing structure that remains true given a more limited scope of parameters (e.g. the advent of Relativity to Newtonian physics or Quantum physics to classical mechanics). The traditional scope of astronomical theory might remain unaltered given the state of knowledge of the solar system that existed 35 years ago. But given what is known today, there is no way that it can be maintained as true and viable.

The case of Saturn says it all. Certainly NASA has assimilated this data and can present their entourage of ad hoc after the fact explanations, but this is only belaboring the hard realization that must occur before real progress can be made. The Axioms of Astronomy must give way to the ***ELECTRICAL NATURE*** of the solar system. Attempting to continue giving "magnetic" arguments without recognizing the underlying cause (dynamic electric fields and electric plasma currents) is a folly that will lead to more stagnation and lack of progress.

With this as an introduction, the two additional papers are included here ... the "Galaxy" paper and the "Saturn's Sweeper Moons Predicted" paper, both published in 1980.

One last note on the Saturn Paper. It was first submitted to the Cornell University Space Science journal ICARUS in preparation for the Voyager I passage of Saturn. I received a short curt response from then editor Joseph Burns who rejected the article from being published stating, "The workings of the rings of Saturn are well understood and are comprised of small particles left over from the formation of the solar system 4.5 billion years ago. There is no need to introduce electrical concepts to explain their properties" (I still have his letter in my files).

I resubmitted the article to the journal The Moon and the Planets where it was immediately accepted for publication without contest. It is most interesting to note that Dr. Burns went on to build his career and became NASA's "expert" on the "Sweeper Moons of Saturn". He even published a book on the topic including the electrical nature of the Saturnian rings, moons and spokes that were mysteriously found as Voyager I passed the following year. Besides the "electrical" nature of the rings, they were found to spiral inwards and therefore contrary to the Joe Burns "traditional theory" statement, the ring particles could not possibly had been left over from any time in the distant past.

So not only was he wrong in rejecting my paper on every count that he made, but he then had the nerve to steal the material *including the terminology* and built the main topic of his career on the paper that he rejected. It is well known that he amongst others at Cornell were instrumental in my being removed from my two teaching positions there (because they 'do not allow incorrect physics to be taught here') ... so much for honesty and integrity in the sciences.

These two articles are included below. For most readers, a reading of the text only will illustrate the results and you can ignore the math and technical jargon. The real purpose here is to place them before the 3 Part Comet Paper (Appendix 2) as these are in the historical order in which they were published. You will see the headers of the 2 articles in this Appendix list my credits as the Mathematics Department of Cornell University. Although they were written before I was fired from the Cornell Physics Department, they were not published until I was rehired into the Math Department.

As a final note, I will recount the meeting I had with Professor Fuches, then Chairman of the Mathematics Department of Cornell, at the end of my second year in the Mathematics Department. He called me into his office and I sat down. He told me he was being pressured by the Astronomy and Physics Departments to fire me (as they had). They were angry that my articles were being published. His exact words were, " I'm not sure what to do here. Your teaching is excellent. I just had 350 students come to me with a petition and request that you take over their Calculus class besides the one you are presently teaching. You are publishing refereed papers in the Space Science and Astrophysics journals, something no one has ever done in the Mathematics Department. You are well liked by your peers in Mathematics and the journal editors who are publishing your papers speak highly of your innovative work."

Two weeks later my contract was not renewed and I spent the last two weeks talking with professors around campus. Some clearly did not like me, while others stood up and shook my hand as they had seen this happen to others there in the past.

CONTINUING GALACTIC FORMATION: A NEW CONCEPT OF GALACTIC EVOLUTION

J. M. McCANNEY

Department of Mathematics, Cornell University, Ithaca, N.Y., U.S.A.

(Received 2 May, 1980)

Abstract. A new theory for galactic arm formation shows the arms to be continually eminating from the galactic nucleus due to a continual influx of cosmic dust. In the neighborhood of the nucleus the problem is treated as a fluid flow and a simple solution is given using conservation of momentum. When rotational dynamics are included the spinning arm system is the result. This solution resolves the problem of the 'missing mass', accounts for warped disk galaxies and gives a probable source for the gravity waves measured by Weber which eminate from our galactic center. Reversal of arm direction is demonstrated and examples of such reversals are cited. An approximate theoretical estimate of the age of our Sun is found to be in good agreement with radio isotope dating. A general result shows why twin star systems are in such great abundance in a galaxy. It gives a model of galactic evolution which begins with only a single massive nucleus with the collapsing gas clouds forming the arms.

1. Galactic Arm Formation

Figure 1 illustrates a homogeneous cloud of cosmic dust and gasses as it collapses around a hemisphere which represents a half model of a galactic nucleus. The matter is considered to be in a fluid state in the vicinity of the nucleus. Viscosity between the fluid and the nucleus is neglected at this point but evidence will be presented later which indicates that the angular momentum of the nucleus is instrumental in maintaining the galactic plane. The flow problem is

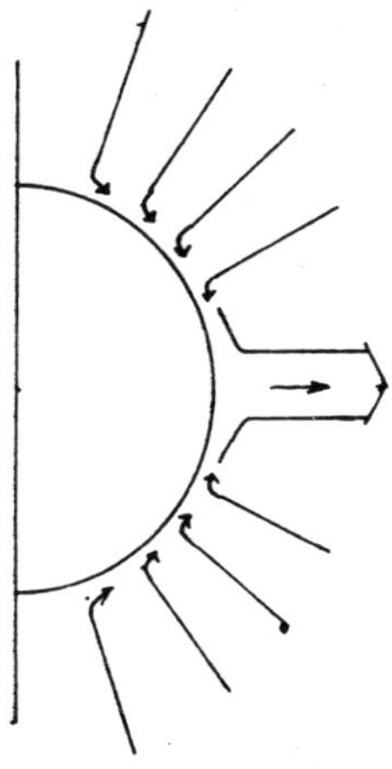

Fig. 1. Flow of gas clouds collapsing around a half-model of the galactic nucleus.

Astrophysics and Space Science **74** (1981) 57–64. 0004–640X/81/0741–0057$01.20.

divided into three distinct areas: (1) cloud influx, (2) fluid flow in the vicinity of the nucleus, and (3) a 'shotgun' effect as the matter is sprayed outwards from the nucleus and condenses into new stars. The analysis of the shotgun effect in producing many stars in an infinite variety of sizes will show why twin star systems are in such abundance. The hot stellar matter sprays outward in Keplerian elliptical orbits ($\epsilon = 0.75$ used in this article) shown in Figure 2.

Due to the back pressure caused by the constant inflow of matter, a single out-flowing arm of circular cross-section will result which moves perpendicular to the base of the hemisphere. This circular cross-section allows the least amount of surface area between inflow and outflow (i.e., lowest surface viscosity). The same quantity must flow out as flows inward. It can easily be shown that if two outflowing arms resulted, the resistive viscosity between the two arms and the inflow would be $\sqrt{2}$ times greater than the case of one flow. (In general total viscosity is proportional to $\sqrt{n}$ where n is the number of outflows.) The arm will choose the path of least resistance. Therefore, a single arm of circular cross-section is forced out of the galactic nucleus. It will be perpendicular to the plane of the base of the hemisphere. As a thin hemispherical shell of dust falls onto the hemisphere the galactic nucleus will be forced to recoil perpendicular to the base of the hemisphere using conservation of momentum. Thus, the arm must eminate in the opposite direction being also perpendicular to the base of the hemisphere. Putting two such hemispheres together gives a model for the arm generating process. In the real galaxy the faint gas clouds which are continually falling inwards are not homogeneous and will fall to the nucleus with a certain angular momentum, thus defining the plane of the Galaxy. This is the cause of the spinning arm system. The matter in the arm is thrown outwards in elliptical orbits with high eccentricity. The total energy is designated by initial conditions. This article assumes the following parameters but they are not meant to typify any particular galaxy. (The model can, however, be used to duplicate the past movements of a galaxy and tell much of its history of encounters with Magellanic clouds. A few examples are presented in this article but a complete analysis should be done. Most galaxies show tremendous symmetries even to very small detail when viewed by this model.)

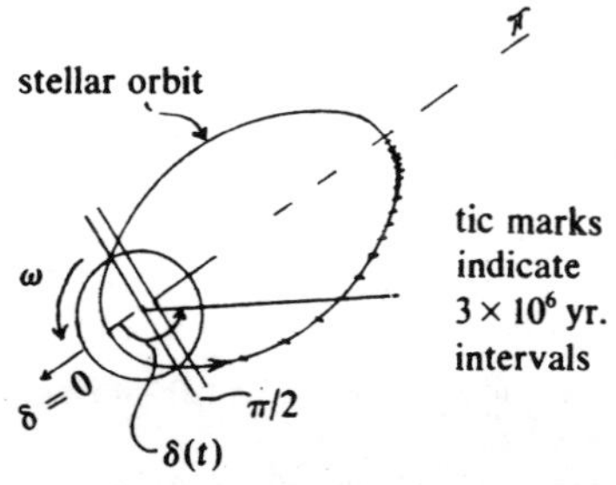

Fig. 2. The orbit of a single star in the Galaxy.

$\epsilon = 0.75$
R_g = Galactic radius $= 10^{20}$ m
a = semi-major axis of ellipse $\simeq 4/7 \times 10^{20}$ m
$b = 0.6a$
R_n = radius of galactic nucleus $\simeq 1/10\ R$ gal.
M_g = mass of galaxy $= 10^{41}$ kg.

It is assumed that about one-third of the galactic mass is in the nuclear region and that the angular velocity of the nucleus is a constant. This idealized case is illustrated in Figure 3. The ellipses are rotated backwards in time to locate where the stars are that left the nucleus at that time in history. This has been done numerically and a typical spiral arm shape is duplicated (see Figure 3). In this example the galactic nucleus rotates 200° in the time it takes a single star to move from the surface of the nucleus ($\pi/2$) to 'aphelion' which is 45 million years. The ratio between these two times is the 'spiral constant' and varies from between near zero to as much as 5 in real galaxies. A value of 5 is a galaxy whose visible arms wind five times around the nucleus. This value also changes with time in real galaxies, depending on the angular momentum of the infalling cosmic clouds.

From Kepler's second law, it follows that

$$\frac{A_T}{P} = \frac{\text{area of ellipse}}{\text{period of rotation}} = \text{constant} = \sqrt{\frac{MGb^2}{4a}}$$

for the total ellipse or any triangular subdivision with the galactic nucleus at one

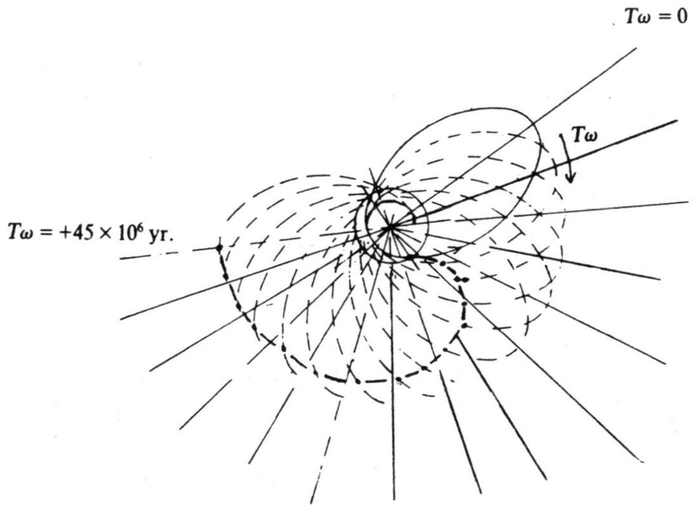

Fig. 3. The present positions of stars are plotted. Ellipses rotate backwards in time to plot the spiral arm shape.

focus. The differential form gives

$$\frac{dA}{dt} = \text{const} = \tfrac{1}{2}r^2 \frac{d\delta}{dt}, \tag{1}$$

where

$$r = \frac{a(1-\epsilon^2)}{1+\cos\delta}; \tag{2}$$

$$\Delta T = \left[\frac{a^2(1-\epsilon^2)^2}{2K}\right] \frac{\Delta\delta}{(1+\cos\delta)^2}; \tag{3}$$

$$\Delta T = \frac{a^2(1-\epsilon^2)^2}{\sqrt{MGb^2/a}} \left(\frac{\Delta\delta}{(1+\epsilon\cos\delta)^2}\right); \tag{4}$$

$$T = \frac{1}{K}\int_{\sim\pi/2}^{\delta} \frac{d\alpha}{(1+\cos\alpha)^2}, \quad \frac{1}{K} = 18.15\times 10^{-15}; \tag{5}$$

$$T_{\text{Tot}} = \frac{1}{K}\int_{\pi/2}^{\pi} \frac{d\alpha}{(1+\cos\alpha)^2} \cong 45\times 10^6 \text{ yr}; \tag{6}$$

where T is the total time for a star to go from the surface of the galactic nucleus to δ (α is a dummy) and T_{Tot} is time to pass from the galactic nucleus to aphelion. Since the galactic nucleus has rotated at a constant angular velocity ω throughout this time,

$$\omega = \frac{200^\circ\,\pi}{180}\,\frac{1}{45\times 10^6 \text{ yr}} = 4\times 10^{-6} \text{ deg yr}^{-1}, \tag{7}$$

$$\omega T = \frac{200^\circ\,\pi}{180}\,\frac{1}{45\times 10^6 \text{ yr}}\,\frac{1}{K}\int_{\pi/2}^{\delta} \frac{d\alpha}{(1+\cos\alpha)^2}, \tag{8}$$

where ωT gives the angle backwards in time that the ellipse must rotate to place a particular star in its present position. Two equations can be solved numerically to give the absolute coordinates of the spiral arm curve at a given point in time.

$$r = \frac{a(1-\epsilon^2)}{(1+\epsilon\cos\delta)}; \tag{9}$$

$$\theta = \delta - \frac{200\pi}{180K\; 45\times 10^6}\int_{\pi/2}^{\delta} \frac{d\alpha}{(1+\epsilon\cos\alpha)^2}; \tag{10}$$

$$90^\circ < \delta \leqslant 270^\circ. \tag{11}$$

The quantity δ ranges from where the star leaves the nuclear surface to where it returns billions of years later after completing its elliptical orbit. The perihelion of the orbit lies within the galactic nucleus thus the star never traverses this portion of the orbit. Figure 3 illustrates one arm of a galaxy in which a particular star moves only 90° in its elliptical orbit while the nucleus rotates 200°. Figure 4 shows two spirals ((a) and (b)) with different spiral const. and a reversal (c). The

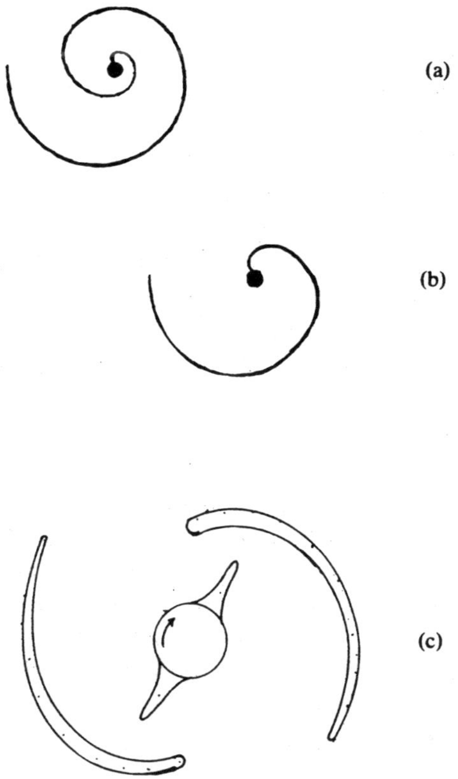

Fig. 4. Spiral arms with spiral constant 2.0 (a) and 1.0 (b). Part (c) illustrates a reversal.

stars normally burn out during their millions of years drifting near their aphelion points and are normally not visible during their return trip to the nucleus.

2. Conceptual Results of Model

The newly formed arm initially has higher angular velocity than the nucleus giving the appearance that the arm is preceding the point on the nucleus where the arm emanates. This area is rarely visible due to the glow of the clouds in the area. In this area are forming giant blue stars and evidently the entire Main Sequence. It seems possible that very small stars may form also and can burn H_2 for a short time as the star is forming in the gravitational field of the galactic nucleus, and not solely by its own mass.

Since the visible galaxy is about 45 million years old, it must regenerate one 45 millionth of this each year. If the Galaxy contains 10^{10} stars then it is producing about 200 stars per year or at a rate of approximately one per day. By the same reasoning, approximately one star per day is returning to the galactic nucleus and it is the collapse of these old stars as they fall onto the nucleus that is a probable cause of the gravity waves being detected by Dr Weber of Maryland. For galaxies with spiral constant greater than a half, the old stars will fall in over the newly forming stars in the arm. This crossing over also occurs in galaxies which have reversed their directions of arm propagation such as in NGC 2523 which is a barred galaxy. The barred galaxies derive their form from an arm which stopped rotating and began propagating with no angular velocity. The arm falls back into itself. Most barred galaxies show a gap at the ends of the bar.

So only about $200M_{\odot}$ of dust clouds fall on the galactic nucleus per year. The globular galaxies seem to be receiving great influxes of clouds which scatter light rendering the entire galaxy invisible. These will be expected to develop massive bright arm systems. The time–scale on which a galaxy works is seen by the estimate of $\omega(=4\times10^{-6}\,\mathrm{deg\,yr^{-1}})$ and the time for a star to pass the first one percent of its distance to aphelion (~30,000 years). In the present crude example, time T_{Tot} (45 million years) is close to the estimated age of our solar system. Slight adjustments in orbital parameters will account for this.

Fusion of the heavy radio-isotopes will occur in the intense star-like environment of the galactic core. This must have important consequences in the future development of solar systems in the Galaxy. The energy which supports this fusion comes from the difference in kinetic energy between the infalling clouds and the outward moving arm (i.e., arms do not flow back to ∞).

Warped disc galaxies (Wright, 1979) can be explained by assuming that the galactic nucleus maintains a substantial angular momentum and does interact with the collapsing cloud flow. The randomly approaching gas clouds will introduce random changes in total angular momentum. If the nucleus contributed no angular momentum, new arms could develop at large angles to old arms. Instead it is observed that only slight changes occur in the plane of the generating arms, causing the disc to 'warp'. Elliptical galaxies appear to be new galaxies in the initial state of formation with some beginning to develop arms whereas the exploding galaxy marks the end of the organized existance of the Galaxy.

The problem of the missing mass (Salpeter, 1977) is also resolved in the present theory. The original galactic theories predicted circular Keplerian orbits. Figure 5(a) shows the predicted orbital velocities. Figure 5(b) shows the v versus r graph for the 'wave theory' which claims the arms to be continually rotating about the nucleus. Figure 5(c) shows the predicted orbital velocities of the present theory as seen by a red-shift observer who sees only the receding component of the arm velocity.

In support of the present theory, a summary of current observations of the Milky Way is given (Geballe, 1979; Wynn–Williams, 1979):

- There are tremendous quantities of radio signals emitted by the Milky Way's nucleus.
- The immediate environment around the nucleus is of hot gaseous clouds surrounding many hot blue stars.

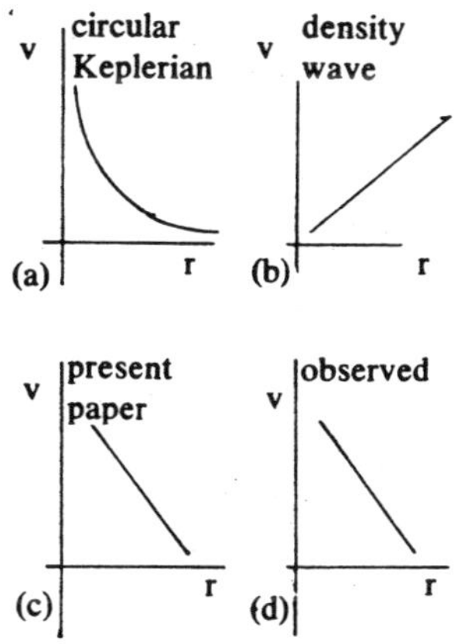

Fig. 5. Predicted radial velocity in special arms for (a) circular Keplerian orbits, (b) density wave theory, (c) present theory, and (d) observed from red shift data.

- Novae and Supernovae are frequent.
- Observation of supergiant stars of short life times indicates that stars are being formed in the galactic center.
- There is also an abundance of old stars near the galactic center.
- The nucleus is an extended object emitting intense radio noise and is about 16 AU in size.
- Gravitational waves are detected eminating from our galactic center at the rate of about one per day.

The twin star system is the result of the shotgun effect as the condensing stars of all sizes are thrown out at approximately the same velocities. Two stars can begin orbiting one another as they leave the gravitational dominance of the galactic nucleus and can remain as a pair until they return to the galactic nucleus billions of years later. Jupiter and the Sun are such a pair and it should be considered that in conjunction with the few stars that are not in twin systems there will most likely reside an unlit companion. It is easy to see how a twin star forms in this way, but why should there be mainly twin star systems and not systems of three or four more stars? Only rarely are even triple star systems observed. Alpha-Centuri is a triple star for example, and has two close orbiting central stars with a smaller relatively distant star orbiting these two. The first two are little affected by the third and the third sees the other two as one. So it behaves much like a double star. The answer is that the shotgun problem introduces only unstable stellar systems into the galactic arm. As a massive newly formed star begins to control the smaller condensed objects around it, they will be distributed on all sides and will take orbits of random eccentricities and orbital planes. Within the first orbit there is a great many encounters with the smallest bodies leaving the system and only the most massive, which generally has an eccentric orbit, remains. Studies have been undertaken to determine the stability of systems with such random initial conditions

(Szebehely, 1974). Results indicate that: (1) randomly selected initial conditions and masses lead to eventual disruption of the many bodied system with the smaller bodies being ejected, leaving in most cases, the two largest to orbit one another. (2) Triple systems are inherently unstable and are therefore rare. Higher order systems are possible but are exceedingly rare. Observation that 75% of the stars exist in double star systems with another 24.9% residing alone, is proof of this statement. As mentioned before, the lone stars will be in most cases accompanied by an unlit companion.

A number of difficulties arise in presently accepted theories of solar system evolution which are not compatible with this concept of galactic evolution. One is that an entire solar system of nine planets and over 30 moons is not the result of a solitary gas cloud evolving on its own but in fact, no more than the Sun and Jupiter could have been in the original system. The other 40 bodies could only have been the object of subsequent captures by this twin star system. Secondly, there is no chance that comets made of condensed ices could form in the heated environment of the galactic nucleus. It is generally accepted in the plasma theories of solar system evolution that the comets were an original part of the solar system (Delsemme, 1972). Two possibilities exist: (1) comets were not an original part of the solar system or, (2) comets are not icy balls or both.

I have proposed (McCanney, 1979a, b, c) a complete theory for comet behavior and solar system evolution based on a new model for the comet which shows it to be a hot radioactive nuclear conglomerate. These papers were not accepted for publication as they did not fit into the accepted theoretical scheme of solar system evolution. This paper lays a firmer foundation for those papers.

3. Conclusion

A new model for galactic evolution was given which predicts the spiral arm shape. Stars are found to be in highly elliptical orbits after being thrown out of the nucleus. The theory correlates heretofore anomalous data and answers the question of 'where is the missing mass'. The most important result is that presently accepted theories of solar system evolution are not compatible with this model.

References

Delsemme, A. H.: 1972, in H. Reeves (ed.), *The Origin of the Solar System*, CNRS, Paris.

Geballe, T. R.: 1979, *Sci. Am.* **241**, 52.

McCanney, J. M.: 1979a, 'Comet Nuclei: An Explanation and Predicted Effects', unpublished.

McCanney, J. M.: 1979b, 'The Role of Comets and Selection Rules in the Formation of Solar Systems', unpublished.

McCanney, J. M.: 1979c, 'A New Self-Consistent Theory of Solar System Evolution', unpublished.

Salpeter, E. E.: 1977, *Ann. N.Y. Acad. Sci.* **302**, 681.

Szebehely, V.: 1974, L. N. Mavridis (ed.), in *Stars and the Milky Way System*, Vol. 2, Springer–Verlag, Berlin, p. 273.

Wright, M. C. H.: 1979, *Astrophys. J.* **233**, 35.

Wynn-Williams, C. G.: 1979, *Mercury-J. Astrophys. Soc. Pacific* **8**, 97.

SATURN'S SWEEPER MOONS PREDICTED

J. M. McCANNEY
Department of Mathematics, Cornell University, Ithaca, New York, U.S.A.

(Received 7 October, 1980)

Abstract. The recent observation of the absorption of radiation belts in the vicinity of Saturn's bright rings and historical observations of the ring system make the following related results apparent:

– The gaps in the rings are caused by the presence of at least 6 small, extremely dense and probably electrically charged 'sweeper' moons which effectively sweep the ring matter clean from the gaps. This is known due to the fading of the inner ring edges whereas the outer edges are well defined. Their orbital periods will differ from the expected Keplerian periods if the moons and Saturn do possess electric fields.

– Absorption of radiation belts near the rings (of Jupiter also) implies that the ring particles themselves are not absorbing the radiation but the small moons are. This is consistent with the observed radiation belt absorption near the outer Saturnian moons.

– If electric fields of the sweeper moons cause the ring edge fading as observed (and not simply gravitational), then Saturn itself must maintain an electric field in its vicinity by way of a sizeable proton wind to affect the uneven ring edge fading and will be surrounded by an H^+ cloud at least to approximately the A-ring. This is consistent with the detection of an H^+ cloud surrounding Saturn (Weiser *et al.*, 1977, p. 755). The other possibility is that these moons are extremely dense and have very large internal magnetic fields.

– Because of their location, these moons must be captured and if very dense as believed, may be core remnants of a nova.

1. Discussion

The possibility of the existence of small moons orbiting Saturn within the gaps of the bright rings has not received attention other than in passing comments by many investigators. Resonance theorists claim that nothing can exist in the gaps and others believe the rings to be composed of material that could not form a moon 4.5 billion years ago inside Saturn's roche limit. Pioneer 11 was unable to confirm the existence of the 'sweeper' moons (except for 1979–S2) because of 'anomalous' data (Filius *et al.*, 1980, p. 425), and no logical explanation for their existence has been previously put forth.

Historic observations (Proctor, 1865, pp. 49 and 61–64; Alexander, 1962, pp. 140–141 and 168) show that the bright rings are sharply defined on the outer edges but fade gradually on the inner edges. This fading is uneven only in the B-ring. These observations also have shown that thin divisions have formed in the rings that later disappeared. These sightings have been ignored since no explanation has been put forth. Pioneer Saturn also detected the uneven fading of the rings (Gehrels *et al.*, 1980, p. 436). Voyager's photograph of the Jovian Sulfur ring also showed this same fading of the inner ring edge (Smith *et al.*, 1979a, p. 933, Figure 10) with the outer edge being sharply defined as in the Saturnian rings. Like Saturn's F-ring, the sulfur ring has a small (~ 40 km diameter) moon skirting the outer ridge edge.

The Moon and the Planets **24** (1981) 349–353. 0165–0807/81/0243–0349$00.75.

The detection by Earth based radio telescope of an atomic hydrogen cloud surrounding Saturn (Filius *et al.*, 1980; p. 415) suggests the presence of a strong proton wind eminating from Saturn. This wind would be similar to the solar wind which seems also to have an excess current of protons. This allows a non-uniform electric field to exist between the negatively charged planet and neutralizing H^+ cloud. The total negative charge of the planet would be equal to the combined positive charge in the proton wind and H^+ cloud. Unfortunately, it is impossible for Pioneer or Voyager to discern if the origin of certain protons are from Saturn or otherwise, however, there are indirect methods (to be discussed) that will implicate their presence. This suggests that Saturn's atmosphere is operating similar to the solar atmosphere but on a small scale.*

Pioneer Saturn's encounter with 1979–S2 showed it to have a very high mass for its size and a strong magnetic field. There are two possible explanations. One is that the small moon was very massive and possessed a very strong internal magnetic field for its size. It can also be explained as due to a large net charge on the moon. A metal space craft in the non-uniform electric field of the moon would feel a strong induced dipole force (i.e., the induced dipole will always produce an attractive force, indicating to telemetry the presence of a 'large mass'). The high relative velocity between the space craft and the charged moon would also induce the high magnetic field at the space craft. As specific data is not available on 'mass' or magnetic field of this moon, a hypothetical example will illustrate the point.

For a cylindrical metal space craft in the non-uniform electric field of a highly charged small moon the ratio of electrical induced dipole force to gravitational force (axis of symmetry of cylinder in the direction of the moon) is given by

$$\frac{F_e}{F_q} = \frac{2q_0^2 \epsilon_0 \pi a^2 \left(1 - \dfrac{r}{r + \Delta r}\right)}{(4\pi\epsilon_0)^2 G M_1 M_2 (r^2)},$$

where q_0 = charge on moon, a = diameter of cylinder = 10 m, M_1 = mass of moon = 2.2×10^{17} kg, M_2 = mass of space craft = 250 kg, r = separation of M_1 and M_2 = 100 km, and Δr = height of cylinder = 2 m.

It is found that $F_e/F_q = 10$ when $q_0 = 300$ coul, corresponding to a surface potential of about 100 MeV on a 200 km diameter moon when the moon has an earthlike density.

Other evidence exists for the presence of small highly charged objects in interplanetary space, although most investigators would rather find alternate explanations for the phenomena. The surface ring formations found concentric around small impact basins on Callisto (Smith *et al.*, 1979b, p. 970) are examples. They are identical to the patterns made by ferromagnetic filings around a current carrying wire (Halliday and Resnick,

* This also supports my contention that the solar interior is not one of collapsed hydrogen, but has a large planetary type core since stars and planets alike have their initial formation in the galactic center (McCanney; 1981). I also believe the solar wind supports a similar H^+ cloud beyond the orbit of Jupiter and that Saturn passes through this cloud constantly as it orbits the Sun, accounting for Saturn's vast atmosphere and appearing as a small planetary nebula.

1965); note that the rings do not begin for a good distance around the impact basin. The small high impact basin and the concentric surface ring patterns can be explained by the approach and impact of a small highly charged object. This implies that Callisto's surface is probably rich in a ferromagnetic dust which formed ringed patterns in the magnetic field of the high velocity impacting object which was charged as it passed through Jupiter's radiation belts.

Space craft charging in a hot plasma has been known for some time (DeForest, 1972, p. 651) and remains a possible source of error in Pioneer and Voyager data (Filius *et al.*, 1980, p. 428). This effect is most noticeable when sunlight is weak as photoelectrons tend to cancel the effect. The complete absence of radiation belts near Saturn's and Jupiter's moons indicates that a similar charging process may occur for a moon passing through a hot plasma with a minimum of reduction due to photoelectrons. A very important point here is that the space craft always charges to a negative potential and thus so would the moons. This is consistent with the inner ring edge fading (i.e., Saturn and the moon must possess the same sign charge as explained below). The mechanism of charging is not well understood, but if one assumes that size is important, and noting that a small space craft can easily charge to 10 000 V, extremely high voltages on a small moon would develop. The inclusion of charged bodies in celestial mechanics has not been considered previously. The result is that the charge to mass ratio is the important quantity to consider. Charge to mass ratios typically are in reverse order of mass, that is, an ion's e/m is very high, that of a ring particle is smaller, e/m of a small moon is still smaller and so on, whereas the total charge on each increases with increasing mass. For this reason, the most noticeable effects occur for only the smallest bodies.

In a previous paper (McCanney, unpublished) the fading of the inner ring edges was explained to be due to the presence of 6 small, very dense 'sweeper' moons which must exist in the gaps. The ring edge fading is caused by an unconsidered effect in celestial mechanics: that the combined force fields of two objects, one more massive (or with greater charge of similar sign) than the other creates a shape in the form of a comet around the smaller object, that is, long on the 'tail' side and short on the head size. It is therefore this combined field that causes the inner ring edges to fade and the outer ring edges to be sharply defined when rotational dynamics are applied (assuming a very massive moon). This same effect was once thought to have caused Jupiter to effectively sweep all the asteroids clear to the orbit of Saturn while allowing them to exist in great numbers in the inside of Jupiter's orbit (See, 1910, p. 193). In the original paper it was assumed that the sweeper moons must be extremely dense because: (1) they have never been observed from Earth, indicating that their diameters must be less than 200 km, and (2) to affect the ring fading to the distances observed (5000 km in the B-ring and 6000 km in the Jovian sulfur ring) the masses would have to be quite high, meaning that the moons would be denser than the heaviest elements known. As this seems unreasonable and for reasons already given, the possibility is raised that these moons maintain a net negative charge as they absorb the radiation belts near the rings. If their orbital periods differ from the expected Keplerian periods, it will be the first concrete evidence of sustained electric fields in the cosmos (that of the moons and of Saturn itself).

The six predicted moons are as follows:

No. 1 lies in the plane of the rings, skirting Saturn's upper atmosphere at closest approach and skirting the inside edge of the crepe ring.

No. 2's orbit lies far ($> 30°$) out of the plane of the rings and partially sweeps out the faint crepe ring. This same moon accounts for the observed non-uniform fading on the B-ring edge. (It fades only at two spots on opposite sides of Saturn.) This moon's orbit crosses the plane of the rings where the fading is maximum.

No. 3 lies in the Cassini division, skirting the B-ring and inner A-ring edges.

No. 4, No. 5 and No. 6 lie in the Encke gap, pioneer gap and at the outer edge of the F-ring, respectively. (No. 6 is 1979–S2).

Saturn's ring system must have a renewable source of ring material which provides the steady-state appearance of the rings. This can come either from the H^+ nebula or Saturn's atmosphere. As many parameters are unknown, meaningful calculations cannot be performed at this time. (The zodiacal disc may be of a similar nature.)

Historic observations have indicated that gaps have appeared in the rings which later disappeared (Proctor, 1865, pp. 49 and 61–64). This can now be reconciled with the fact that Saturn is a capturer of the same objects that constitute the small moons (Cline, 1979) and that Saturn has a renewable source of ring material. Such an object passing through Saturn's system could quickly charge and sweep out a single path in the rings that would later refill after the object was gone.

If the sweeper moons are non-Keplerian (implying a proton wind for Saturn) it will suggest that an asteroid type nucleus moving at high speed through the solar wind will charge also, causing the characteristic comet shape to form around it as + ions are drawn in under forces that can be billions of times stronger than gravitational forces alone. The implications of this model of comet behavior have been worked out completely (McCanney, c) and numerous observed characteristics are easily explained such as 'wandering', sunward spikes, a shrinkage of the coma as the comet approaches the Sun, curved tails, spiralling of tail material, etc. This model shows the comet to be accumulating matter and ultimately may be captured as a planet, asteroid or moon. The detection of non-Keplerian moons around Saturn will support this model of the comet.

2. Conclusion

The existence of at least 6 small dense and/or charged moons are predicted as the sweepers of the Saturnian ring gaps due to historic and recent observations. The detection of non-Keplerian moons in the gaps will imply the existence of a proton wind and associated H^+ cloud in the vicinity of Saturn, providing the first evidence that sustained electric fields do exist in the cosmos. The concept of an asteroid-like nucleus moving at high speed through the solar wind (causing the nucleus to charge) as a plausible explanation for comet behavior will be supported by the existence of the predicted sweeper moons, as they exhibit the same characteristics.

References

Alexander, A. F.: 1962, *The Planet Saturn*, Faber and Faber, London.

Cline, J. K.: 1979, *Celestial Mech.* **19**, (No. 4).

DeForest, S. E.: 1972, *J. Geophys. Res.* **77**.

Filius, W., Ip, W. H., and McIlwain, C. E.: 1980, *Science* **207**.

Gehrels, T., *et al.*: 1980, *Science* **207**.

Halliday, D. and Resnick, R.: 1965, *Physics*, John Wiley and Sons, New York, pp. 753.

McCanney, J. M.: 1981, Continuing Galactic formation. *Astrophys. Space Sci.*

Proctor, R. A.: 1865, *Saturn and its System*, Longman, Green, Longman, Roberts and Green, London.

See, T. J. J.: 1910, *Researches on the Evolution of the Stellar Systems. Vol. II*, Thomas P. Nichols and Sons, Lynn, Mass.

Smith, B. A. *et al.*: 1979a, *Science* **206**.

Smith, B. A. *et al.*: 1979b, *Science* **204**.

Weiser, H., Vitz, R. C., and Moos, H. W.: 1977, *Science* **197**.

XV. APPENDIX 2 - THE 3 PART COMET PAPER

The following is 61 pages of detailed reading that will keep most people reading and researching for some time. The point of including it here is that it was written to be published in the astrophysics journals originally in 1980 following the 2 papers in APPENDIX 1. The original papers did not include any mention of Velikovsky, since I did not know who he was at the time. Other than that, the papers are almost exactly the originals in content and were derived from my first book of over 300 pages (copyrighted 1980) entitled "Origin of the Planets – Comet Capture Processes in the Formation of Solar Systems".

The original papers introduced all of the then new concepts in one large paper. A complete rewriting of traditional theory was needed all at one time. Some suggested publishing bits and pieces. But in doing that, none of this would fit or make sense with anything that had ever been published in the "respectable journals" to date, so it only made sense to hit it all at one time.

The picture included on the next page is that taken by a friend and photographer William Boudreau during an intense thunder storm in Minneapolis. I included it as the cover photo on the issue of the journal KRONOS for part II of the 3 part comet paper. It is so striking a photo as it shows the immense power in a single lightning bolt from an average summer thunder storm. The point is that an electrical discharge from a raging large comet passing near Earth would melt this building and leave a gaping hole in the ground. People are just unaware of the magnitude and awesome power of the dramatic Earth Changes that can be imparted by a LARGE comet.

This is why people must take heed and begin to prepare early for any incoming large comets and new planets since as the objects approach there will be little time to prepare … and even the best preparations may fall by the wayside to the tremendous power stored in these cosmic storms from outside our fragile existence. As already noted, the Earth has been ravaged in the past by these cosmic intruders and it will again in the future. The question is not "IF", but only "WHEN" and "TO WHAT DEGREE".

There are many notable predictions of the 3 part comet paper that have been verified by space probe data, however the most telling is the prediction that comets would exhibit x-rays to the sunward side of the comet nucleus. This was one of my main tests to disprove the Dirty Snowball Comet Model and verify the Plasma Discharge Comet Model. That prediction is given in Appendix II of Part II of the 3 Part Comet Paper on pages 84 and 85 (note the page numbers are the original page numbers from the publication so they do not correspond with the rest of the book). Of course when the x-rays were discovered in 1996 by ESO and NASA, and even though they admitted they had no clue that comets should give off x-rays, they never referenced this work in spite of the fact that ALL the members of the Goddard team had my papers laying on their desks (because I called them and sent these articles to them by registered next day airmail).

Photo of a lightning bolt hitting the IDS Tower, Minneapolis during a summer thunderstorm. The flash blew out the light meter on the camera over one mile away. (photo courtesy William Boudreau)

THE NATURE AND ORIGIN OF COMETS AND THE EVOLUTION OF CELESTIAL BODIES

(PART I)

J. M. McCanney

ABSTRACT: This paper provides an alternate theory for comet behavior and shows comets to be planetary, lunar, and asteroidal bodies in their formative stages. It demonstrates that tail matter is attracted *towards* an asteroidal comet nucleus by strong electrical forces. Additionally, two charging mechanisms are identified, both of which produce a net negative charge on the comet nucleus. This is supported by data from recent space probes. Comet wandering, sunward spikes, a shrinkage of the coma as the comet approaches the Sun, curved tails, the gathering and maintenance of meteoroid streams, spiralling of tail material, and the rapid orbital circularization of large newly captured comets are also discussed.

Earlier papers[1, 2, 3] used similar concepts to predict the existence of strong electrical fields in the vicinity of Saturn, showing Saturn and its ring system to be analogous to the Sun and its zodiacal disk. The realization of the proton wind-supported capacitors of Saturn and the Sun led to a number of unexpected theoretical considerations that included 1) the recognition of the charging process used by comets, 2) the postulation of an ion and dust cloud held back by solar wind pressure near the orbit of Jupiter[4] – which is one source of comet tail matter – and 3) a postulated electric dipole red-shift in photons leaving the central star. Still another theoretical result was the possibility of an electrically induced magnetic dynamo powered by a planet spinning inside the orbit of a slightly charged moon. Empirical correlation between moons and magnetic fields has been known for some time,[5] though the wandering of our Moon has remained an unsolved mystery.[6, 7]

An attempt is made to explain solar system formation from the time a newly formed twin star system leaves the galactic center[1] to when it develops its solar system by the capture of comets. The reader's knowledge of planetary encounter and N-body literature is assumed since it is basic to the paper but unreferenced. However, the text by T. J. J. See,[6] which develops the first capture theory for the origin of the solar system (OSS), is indispensable. A major result of this paper is also the quantization of Newtonian space. Finally, the link between planetary formation, geomagnetic reversals, and biological evolution is examined.

PREFACE

This paper was produced during the 1979-80, 1980-81 academic years while the author was a lecturer in the Physics and Mathematics Departments of Cornell University (Ithaca, N. Y.) Only minor grammatical changes have been made for publication and numerous footnotes have been added for clarification. The article is a condensed version of a 450 page manuscript (*Origin of the Planets, Comet Capture Processes in the Formation of Solar Systems,* also by the author) which further develops each aspect of the new comet theory. Although it was never intended, the theory explains Velikovsky's claims of Venus transforming from a comet into a planet and is supported by data from recent space probes.

Since 1982, with the analysis of data from the Pioneer 11/Voyager I/Voyager II missions to the outer planets and the Pioneer Venus/Russian Venera probes, the trend even among established astrophysicists has markedly turned towards catastrophism based on celestial events (these have been mainly variations on the "colliding asteroid" theory). In spite of this trend and a wealth of new data on electromagnetic phenomena, mainstream astrophysicists continue to maintain that gravity is the only force in the cosmos and to support long standing theories such as the Big Bang, the nebular collapse theory for the origin of the solar system, the greenhouse effect, the ice ball comet model, and General Relativity (all of which are shown to contain theoretical inconsistencies in this paper).

As the data arrived from around the solar system, the author witnessed repeated efforts within the space science community (primarily NASA) to ignore the importance of electrical phenomena. If the data did not fit into the established theoretical picture, after-the-fact theories were contrived to force-fit the data, or the data were simply not dealt with at all. It should have been apparent that the data were unquestionably contradictory to any expectations of traditional theory and that a radically new set of self-consistent concepts would be needed.

Part I is the first of a three part series which develops a new theory for comet behavior and solar system evolution. Many may wonder why a new theory is necessary; thus Part I begins with a brief critique of presently "accepted" astronomical theory and is followed by an introduction to the new comet theory.

I. A BRIEF CRITIQUE OF THE ICE BALL COMET MODEL AND NEBULAR THEORY OF THE ORIGIN OF THE SOLAR SYSTEM

Occasionally letters are published which disagree with the ice ball comet model (IBCM) and nebular collapse theory for the origin of the solar system (OSS).[8] Personal experience also indicates that there is a group of scientists and astronomers who do not accept either theory, but do not publish since they do not know what to publish. This has led to the popularized statement that there is universal acceptance of these theories. After the Voyager I Saturn encounter, many began to realize the inability of the nebular theory to explain the data, especially the electrical phenomena and large energy output of Saturn as compared to Jupiter.

All current literature on planet formation assumes the pre-existence of planetesimals which are then shown to agglomerate relatively quickly into planets.[9] The great difficulty with this has always been in showing how the proto-planets form, since only large magellanic clouds can achieve gravitational collapse in theory (in practice, no one has ever witnessed the collapse of any cloud, no matter how vast its size).[10] Also, if these small planetesimals are so difficult to explain, then how did the relatively small frozen comet nuclei form in the primordial nebula? It has always been assumed that this is how it must have been, as is the case with the Oort-cloud which currently is impossible to detect.

Other objections which cannot be ignored are the results of all four Pioneer-Venus probes[11] which detected "more energy being radiated up from the lower atmosphere than enters as sunlight", the faint glow at the surface and atmospheric lightning, not to mention the high concentration of argon-36,[12] among others. At this point science cannot be advanced by simply trying to modify previous theories which fall very far short of explaining these data or by refusing to look at new approaches to the problem.

The unexpected elevated temperature in Titan's clouds[13] has been explained as due to a temperature inversion, suggesting that the heat is generated by a greenhouse effect. But, Titan receives only about 1/40,000th the sunlight that reaches Venus, so few will believe in a greenhouse effect at this distance from the Sun.[14] Infrared data must be viewed skeptically as they have consistently given low temperatures in Earth-based data (*i.e.*, Venus, Jupiter, and Saturn). Also, Pluto is now known to have gaseous methane in its atmosphere and therefore must have a considerable N_2 atmosphere to hold this in place (as with Titan). With the recent determination of the low mass of Pluto (Pluto has non-trivial amounts of gaseous methane in its atmosphere yet is only 1/400th the mass of Earth),[15] one can only ask how it has maintained this atmosphere for 4.5 billion years and how it maintains the elevated temperature necessary to have a gaseous atmosphere (as with Titan)?[16]

The current sheet of five million amps that flows constantly from Io to Jupiter was the first electrical discharge phenomenon recorded by Voyager in interplanetary space. It is generally stated[17] that the auroras on Jupiter arise from current flow from Io's torus. But, since auroral spots – one near each pole – also follow Io as it orbits Jupiter's dark side,[18] the current must be coming from Io itself. The visible auroral spots near Jupiter's poles following Io were mentioned in early news releases but no reference to them was found in the issues of *Science* (written by NASA space scientists) dealing with Voyager I and II's encounter with the Jovian system, *i.e.,* 1 June 1979 and 23 November 1979.

Although a "magnetic" explanation was given for Io's current sheet, current cannot flow unless a potential difference exists; therefore Io must maintain a net charge with respect to Jupiter. This paper will show that Io maintains a net electric charge, using the same charging process as comet nuclei orbiting the Sun. It will be shown that Jupiter's spinning inside the orbit of charged Io creates Jupiter's magnetic field, and not vice versa.[19]

The widely publicized tidal heating of Io to account for its great internal heat and volcanism has been questioned.[20] The tidal theory predicts the greatest heat to be at the north and south poles of Io,[21] but almost all volcanic activity is observed within 30° of its equator. As with all "accepted" theories, it has been favored because it supports the *a priori* assumption that everything in the solar system formed 4.5 billion years ago. Internal heating will be discussed in detail and it will be shown that tidal heating has been overestimated. The heat is rising from Io's young interior through volcanism which is a result of quakes caused by the tidal action of Jupiter, Europa, and Ganymede.

Other important but often ignored anomalies are the wanderings of Neptune and Earth's Moon, the selective heavy cratering of the far side of the Earth's Moon,[22,23] and small halos around certain asteroids.[24]

Most investigators strictly hold that electrically charged celestial bodies cannot exist because it would be observed in the planetary motions. The answer to this is that it is now known that the charge to mass ratios of celestial bodies vary greatly with size, with the most notable effects occurring only in the motions of the smallest bodies. This is discussed at length throughout the present paper and has been observed in Saturn's system.

Although much has been written and many calculations performed on the ice ball comet model, it is difficult to imagine that the miniscule amount of solar radiation falling upon this nucleus can cause comas 1.5×10^6 km in diameter and 100 million km in length.[25] Furthermore, the comet must continually fill this space as the tail follows the comet in its orbit. Piecewise integration suggests that the comet would have to fill this volume at least 600 times during a single passage while inside the orbit of Mars and it is expected to do this on thousands of returns. Also, it is particularly hard to imagine a sungrazing ice ball passing through the 1 million degree solar corona, spending a number of hours grazing the solar atmosphere (not to mention passing through the solar Roche limit on a highly eccentric orbit), and passing to the outside again relatively unaffected.

If the icy volatiles were ejected by solar radiation bombardment, then the statistics of following such a molecule, given the mean free path as a constant, would show the comet coma luminosity to fade exponentially as the distance from the nucleus. On the contrary, the coma is well defined up to an edge and does not exist appreciably beyond this.

Observed anti-tails (sunward fan-shaped tails) have been explained as due to the rotating ice ball interacting with the solar wind.[26] Comet wandering is claimed to be due to the ejection of jets from the ice ball.[27,28]

The curved Type II tails as observed in Donati's comet and comet West follow the comet in its orbit. This would require a selective curvature in the solar wind or other such containment mechanism in the IBCM; however, it has been shown that neither solar wind nor solar radiation can provide such a containment mechanism.[29,30]

The IBCM is only valid for the region of space well within the orbit of Jupiter. The two largest comets in history (comet 1729 and 1927 IV) were recorded outside Jupiter's orbit (1927 IV was seen beyond the orbit of Saturn).[31]

Comets with sunward "spikes" are explained in the IBCM as a thin Type I tail which only "appears" to extend in the sunward direction. This explanation was first given when Comet Arend-Roland developed a sunward spike for seven days during its 1957 passage.[32] Comet Kohoutek developed a similar spike as it passed near the Sun as seen by Sky-Lab astronauts.[33] Since then, other spiked comets have been observed, always occurring in the ecliptic near the Sun.[34] Pliny the Elder in his *Second Book of Natural History* speaks of comets that project horns, and there are many other ancient references to unusual comets. Data which must be included are the 6,000 year old American Indian rock paintings found near Green River, Utah.[35] The paintings are unmistakably those of a comet with a spike in the form of a helix. One shows the comet with a large nucleus, the other without. As the comet is drawn twice, it was the painter's intention to draw the sunward spike in the form of a helix. Spiralling of comet tail material is also commonly observed, reminiscent of ions moving in a magnetic field.

The new comet theory shows the sunward spike to be part of an electrical discharge, and the spiralling of the spike and tail matter a result of charged particles moving in a magnetic field supported by the charged comet nucleus. This is a marked difference in the theories and, therefore, provides one of the many Earth-based experiments that can decide between the two theories. If radio noise is detected during spike formation, then the IBCM cannot explain this since the thin Type I tail should be much less active than the larger Type II tail. A magnetically induced discharge in large Type I tails has been suggested[36] but this cannot be related to sunward spikes for the reason just given. Also, low level radio noise in comets has been accidentally detected during occultation of stars.[37] So detection of excessive radio noise in spiked comets should provide a definitive test for the alternative theories.[38]

The link between galactic and solar system formation is necessary for a complete understanding of celestial phenomena. The traditionally accepted density wave theory of spiral arm formation is consistent with the nebular theory of OSS in that it explains the origin of impulses believed necessary for stellar collapse and formation. There are difficulties which still remain with this model, however. The mechanism that begins the density wave remains unidentified as an interaction of galactic proportions is needed which is common to *all* galaxies. This is further complicated by the need for a symmetric interaction to explain fine detail such as star densities – irregularities in the spiral arm shape and the anti-symmetric warping of the galactic disk – all of which occur with great symmetry even though the arm pairs are separated by distances of hundreds of thousands of light years.

II. COMETS AND THE DISCHARGE OF THE SOLAR CAPACITOR

IIa) *Galactic Formation, Saturn, and the Charging Processes Used by Comets*

Two papers have preceded this one and must be read with their references to interpret the present paper properly.[1, 2]

A new concept of galactic evolution has been proposed in one paper which is consistent with the new comet capture theory of OSS and is contradictory to both the density wave concept and nebular collapse theories of OSS. The new model's main result (as related to the present paper) is that twin star systems are formed near the galactic nucleus.(39) It also explains the cause of the high degree of symmetry found in galactic structure.

The twin star system is necessary for capture in celestial mechanics; thus the formation of binary stars that are observed in abundance in the sky is an essential part of OSS by capture processes. The dynamics of twin star formation in the spiral arm, as it leaves the galactic nucleus, also provides an important source for the asteroidal comet nuclei which can become captured by a twin star system. As the spray of condensing matter leaves the gravitational dominance of the galactic nucleus, the largest conglomerates will begin to control the volume of space around them, with the smaller objects assuming orbits in random planes with random eccentricities about the central more massive star. Within a short time, there will be a great number of encounters.

This leaves, in most cases, the two largest bodies to orbit one another with the smallest bodies being ejected from the system. These smaller bodies are observable as the dispersion of light that occurs around the spiral arm near the galactic nucleus as they move outwards for possible capture by twin star systems. It is a game of numbers; of the multitude of asteroidal planetary "seeds" ejected from this portion of the spiral arm, only a few will eventually become active members of a solar system. Here also it is seen that the planes of the solar systems formed will be randomly oriented as will the orbital directions of the smaller stars of the pairs.(40) Jupiter and the Sun were the original twin stars of our system, with the rest of the planets, moons, and asteroids being captured one by one at a later time, the selection rules being governed by chance.

Some may ask: "why cannot some planets have been part of the original system as it left the galactic nucleus?" If such 3 (or N) body systems were possible when given random initial conditions, then triple star systems (and higher order systems) would be more abundant. Only 0.1% of all stars are in higher order systems, and the known systems (*e.g.*, the triple-star alpha-centauri) act much as a twin star, *i.e.*, a closely spaced binary with a distant orbiting third star.

Comet captures are well understood and well documented, *e.g.*, Lexell's comet 1770 and comet Brooks II, 1886. Also, the observation that 5000 asteroids lie within Jupiter's orbit with only a few beyond shows the organizing effect of capture by a twin star. Thus, with over

75% of the stars in double systems, and many single stars with unlit companions, developing planetary systems should be found in all of these.

The second paper deals with the star-like nature of Saturn (and Jupiter). From the observed central high velocity wind belt, the highly developed ring system (analogous to the zodiacal disk), the electric discharge phenomena in its vicinity and the proportionately larger thermal output when compared to Jupiter, it must be true that Saturn is much more active and therefore younger than Jupiter.(41)

A major result of the Saturn paper is the identification of two charging processes, both resulting in a net negative charge on a body moving in a hot plasma (either planetary radiation belts or the solar wind). The first has been detected and is induced as the body enters regions of varying electrical potential within the plasma.(2,42,43) A small space craft can quickly charge to a potential of 10,000 volts, so if size is assumed to be important, then a small asteroidal body could quickly charge to a substantial voltage.(44) This was observed when Pioneer-Saturn passed under the small asteroidal moon 1979-S2 and experienced a "great mass" with a large magnetic field. The great mass sensed by telemetry was the result of the induced electric dipole force on the metal space craft as discussed previously(2) (and the same force responsible for the gathering and maintenance of meteoroid streams by comets, to be discussed).

The second charging mechanism occurs during the discharge of the Sun's (or Saturn's) capacitor formed by an excess current of protons in its solar wind. The capacitor forms between the negatively charged central star and positively ionized nebular cloud which surrounds the star in the shape of a donut. The discharge of this capacitor is triggered by the intrusion of an already charged asteroidal body(45) (charged initially by the first process). Current flows in a line between the star and surrounding neutralizing ion cloud via the comet nucleus. Electrons flow outward from the negatively charged star (sometimes visible as the sunward spike) while positive ions flow inwards from the nebular ion cloud (forming the comet tail). Due to the higher mobility of electrons, they arrive in greater numbers at the asteroidal comet nucleus, causing a build-up of negative charge on the nucleus. It is the combined electric fields of the Sun and comet nucleus which create the characteristic comet shape (to be discussed). Fan-shaped anti-tails are caused by ions and protons from the solar wind which also pour into the comet nucleus from the sunward side and fluoresce as they recombine with electrons.

The first charging process depends on the size of the object, its velocity relative to the plasma, and the intensity of the radiation belts (or solar wind). The second charging process does not depend on the object as much as on the discharge itself. Once the second process begins, it is self generating until the entire solar capacitor is drained. This can lead to enormous voltages on the comet nucleus. The greater this charge becomes, the greater the discharge becomes, and thus is self-generating. Although it appears that celestial objects in circular orbits have less charging than those in eccentric orbits, what terminates the charging process is still not well understood.[46]

The standard argument raised against the existence of electric fields in space is that "in the plasma environment of space, any charge accumulation would be quickly neutralized". This would be true if this plasma were not controlled by a powerful charge separating cell in its center (the Sun). The mobility of free space charge must be taken into account as it is the varying mobility of charged particles that causes the comet nucleus to charge. Thus the Sun acts as a Van de Graaff generator while the solar wind holds back a nebular ion cloud estimated to begin somewhere near the orbit of Jupiter.[47] This is one source of matter for the comet tail and will be seen to contain the light elements up to at least sulfur. The heavier elements will be seen to come from the other source of tail material, the zodiacal disk.

It has always been assumed that the solar wind contains equal currents of electrons and protons to maintain an electrically neutral solar system. But there is no reason for assuming this is so. It would be impossible to detect the overall current leaving the Sun at any given moment. The few points at which the solar wind has been monitored can in no way be extrapolated to say that there are equal currents of protons and electrons in the solar wind as has been done by theorists. External characteristics, *e.g.*, comet phenomena and electrical phenomena in Saturn's rings, indicate that there must be an excess current of protons. In solar prominences are seen composite streamers of similarly charged particles moving in the local magnetic field, so there can be no doubt that the Sun has the ability to selectively eject composite streamers of similarly charged particles.[48]

IIb) *Comet Theory*

Figure 1 shows the progression of characteristic comet shapes formed by the *combined* electric fields of the Sun (q_1) and comet nucleus (q_2), both with net negative charges ($q_1 > q_2$) as the comet orbits the Sun. The comet shape is defined by the area in which positive ions will be trapped by the comet nucleus due to the *combined* electric fields.[49] Due to its electrical charge, the comet nucleus be-

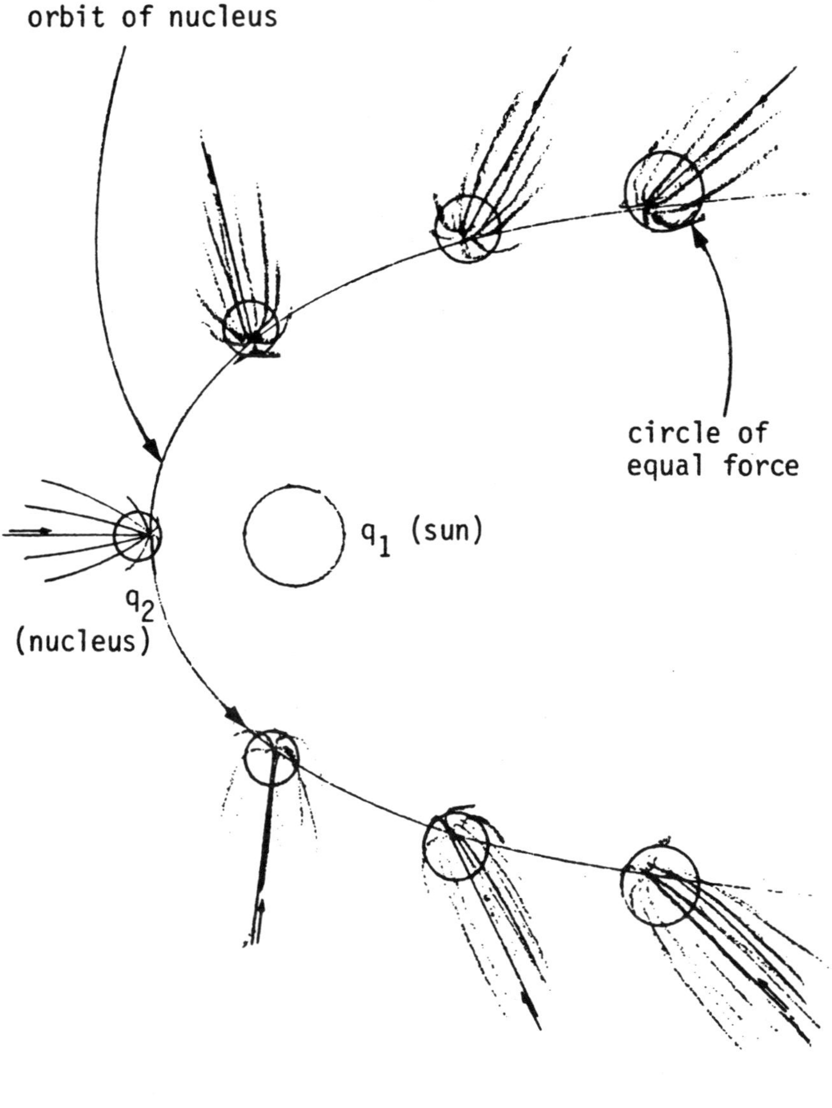

Figure 1: Computer plot of characteristic comet shapes (q_1/q_2 = constant) for a comet following a parabolic orbit. The shrinking of the coma is predicted by theory, showing the CEF as related to the Sun-comet nucleus separation.

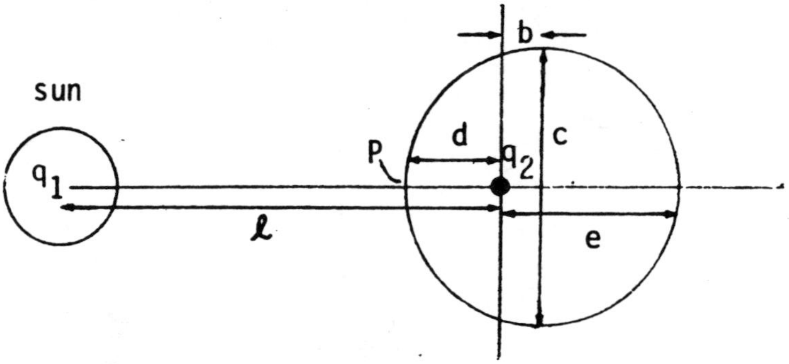

Figure 2: Dimensions of the CEF ($q_1 > q_2$).

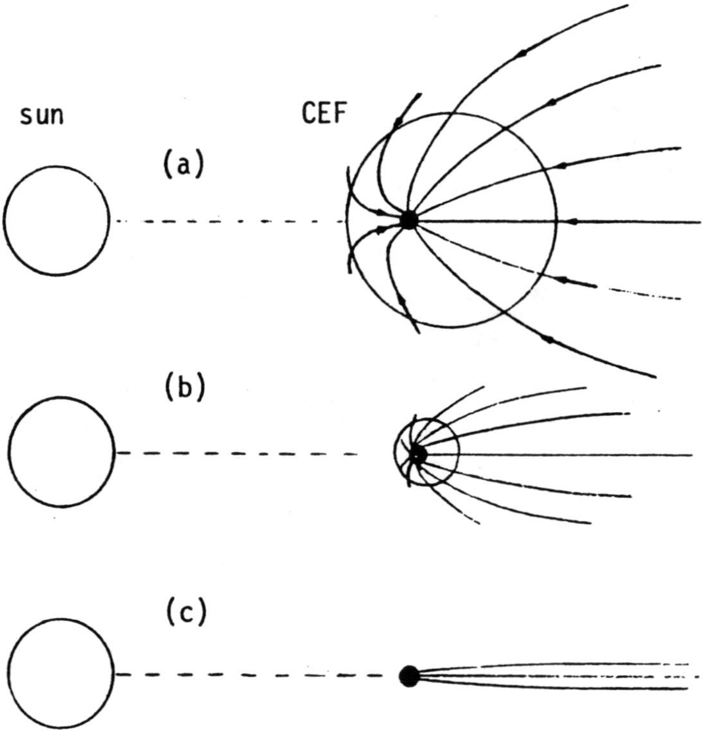

Figure 3: Comet shapes for varying values of $q_1/q_2 = k$.
a) $k = 10$
b) $k = 100$
c) $k = 1000$

comes a singularity in space with a near infinite supply of tail material. The equations governing the size of the coma are given by:

$$b = \frac{\ell}{(k - 1)} \quad (1)$$

$$c = \frac{2\ell\sqrt{k}}{(k - 1)} \quad (2)$$

$$d = \frac{\ell}{\sqrt{k} + 1} \quad (3)$$

$$e = \frac{\ell(1 + \sqrt{k})}{k - 1} \quad (4)$$

where $q_1 \gg q_2$, $k = \frac{q_1}{q_2}$, P is the Lagrange point of the electric field and l is the Sun-comet nucleus separation (*Figure 2*). The circle (diameter = c) called the circle of equal force (CEF) is where the acceleration on an ion towards the Sun is equal to that of the comet nucleus. It is easily shown that the vector sum of these forces always points towards the Lagrange point. The equation of the CEF, with q_2 at (0, 0) is:

$$x^2 + y^2 + \frac{2x\ell}{1 - k} = \frac{-\ell^2}{1 - k^2} \quad (5)$$

(for proofs and illustrations, see *Appendix I* and *II*). The shrinking of the coma as the comet approaches the Sun is accounted for in equation (3) which also provides a simple method for determining the value k from Earth-based measurements. This is one common characteristic of comets which must be explained in theory.[50] *Figure 3* shows varying comet sizes for varying values of k. The stratification sometimes visible in comet tails[51] is caused by variable concentrations of dust and gases in the tail area as they follow field lines into the comet nucleus.

The tail area extends to infinity theoretically but in practice it extends only to the neutralizing nebular ion cloud near the orbit of Jupiter.[52] Thus tails of extreme length (Type I) are seen as sunlight reflected from dust that drains in from the nebular cloud during the electrical discharge of the solar capacitor. Time of flight of tail dust is negligible compared to the movement of the comet nucleus. Therefore, the static case is an excellent approximation to the real situation. The tail material causes a build-up and layering of the comet nucleus and is not melting away as hypothesized in the IBCM. Red shift data are available which are interpreted as indicating the velocities of volatiles moving away from the comet nucleus. Section IV of this paper will show why these data have been misinterpreted.

Electrodynamics in the solar system constitutes a widely misunderstood field of study among traditional astronomers. It is now apparent that charge-to-mass ratio is of great importance in celestial mechanics. This has been demonstrated in Saturn's system where the most noticeable effects occur in the smallest objects. The newly discovered rotating spokes in Saturn's rings have been explained(2) as due to the sweeping of small ring particles that are ionized during the discharge of Saturn's proton-wind-supported capacitor, the area swept out being the "shadow" of the comet tail. This is supported by the location of a bright ringlet from which the spokes always emanate. This ringlet contains orbiting "charge centers" (either small moons or large ring particles) which act exactly like comet nuclei when inside the zodiacal disk of the Sun (a second source of comet tail material).*

The following example illustrates the forces involved in tail production due to varying charge-to-mass ratios, comparing gravitational and electric effects. A comet nucleus (10^{15} kg) at 1 A.U. from the Sun develops a charge sufficient to cause a coma 50,000 km to form (measured in the sunward direction). This requires (by Equation 3) a charge ratio between the Sun and comet nucleus of $q_1/q_2 = k = 10^7$. The gravitational force of the Sun on the comet nucleus is approximately 2×10^7 nt. Table 1 gives values of charge and mass for the Sun (q_1, m_1), comet nucleus (q_2, m_2) and a singly ionized CO molecule (q_3, m_3) assumed to be in the tail area of the comet. Note that the charge-to-mass ratio (q/m) ranges over 31 orders of magnitude.

TABLE 1

	Mass (kg)	Charge (Coul.)	q/m (Coul./kg)
electron	9.1×10^{-31}	-1.6×10^{-19}	-1.76×10^{11}
CO^+ ion	4.7×10^{-26}	$+1.6 \times 10^{-19}$	$+3.3 \times 10^{6}$
comet nucleus	10^{15}	-2.2×10^{3}	-2.2×10^{-12}
Sun	2×10^{30}	-2.2×10^{10}	-1.1×10^{-20}

*[Cf. Earl R. Milton, "Saturn and Voyager", KRONOS VI:3 (1981), pp. 55-62 and M. Brandt and M. Bodner, "Electromagnetic-Gravitational Coupling Phenomena in the Saturn Ring System", KRONOS VI:3, pp. 63-70. – *LMG*]

An all too common remark among traditional astronomers is that, if electrically charged bodies existed in space, we would observe fantastic accelerations in these bodies and the Universal Gravitational Constant would be noticeably altered. The present example shows, however, that the resultant repulsive electrical force between the Sun and comet nucleus is only one part in 10^{12} of the attractive gravitational force. So the orbital perturbation on the comet nucleus will be negligible and the value of the gravitational constant altered by only one ten billionth of one percent. This is due to the low charge-to-mass ratios of the Sun and comet nucleus.

Ions, however, have relatively high charge-to-mass ratios. The ratio of electrical force to gravitational force on a CO^+ ion in the comet tail is given by:

$$\frac{F_e}{F_g} = \frac{\frac{q_3}{4\pi\varepsilon_0}\left(\frac{q_1}{r_1^2} + \frac{q_2}{r_2^2}\right)}{G\, m_3\left(\frac{m_1}{r_1^2} + \frac{m_2}{r_2^2}\right)} \stackrel{\sim}{=} 10^{14}\left(\frac{10^7\, r_2^2 + r_1^2}{6 \times 10^9\, r_2^2 + r_1^2}\right)$$

The resultant attractive electrical force on the CO^+ ion is at least 10 orders of magnitude larger than the gravitational force due to the mass of the Sun and comet nucleus.

The charged comet nucleus is capable of drawing in vast amounts of matter by powerful electrical forces. This occurs without an observable perturbation in the orbit of the comet nucleus. Later it will be shown that the observed "wandering" of comets from their orbits is caused by the drag of the tail on the comet nucleus, electrical perturbations being noticeable only over long periods of time.

Thus, a comet involved in the discharge of the solar capacitor will continue to grow in size and mass. This is why a good deal of radio noise should occur in a well developed comet whereas the IBCM would predict relatively little when compared with known sources such as Saturn's rings.

The criticism has been raised that if such a charge existed on the Sun, then extremely high energy particles would be commonplace in the inner solar system. This, however, again is analogous to the Van de Graaff generator in which the charge need only be sprayed off the central belt as it will assume its state of lowest potential and without excessively high energy particles involved. The *discharge* of the solar capacitor will involve very high energy particles, but not the charging of this capacitor.

There are numerous reasons for identifying the other source of tail matter as the zodiacal disk. This is reinforced by the analogy between Saturn's spokes and comet phenomena near the Sun. Curved tails, such as in Donati's comet when it neared the Sun, are a result of the matter in the rotating zodiacal disk falling into the comet nucleus which is viewed as a singularity in space. (Another cause for other Type II tails will be discussed later.) This is due to the relative motion of the comet nucleus with respect to the zodiacal disk. Donati's comet also exhibited a pair of thin Type I tails which must have arrived from the nebular ion cloud.

Comets begin to show heavy elements (such as nickel, potassium, iron, etc.) in the tail as they approach the Sun.[(53)] These come from the zodiacal disk (iron has been detected in Saturn's rings which is analogous). Thus, a fundamental aspect of planetary formation is the amount of time a comet spends inside the zodiacal disk (where heavy elements accumulate) as opposed to inside the nebular ion cloud which provides only lighter elements up to approximately sulfur, determined from observation of tail ions.[(54)] This fact may account for the abundance of low density celestial objects found past the orbit of Jupiter while still allowing some of them to be of earth-like density. Layering of matter on the comet nucleus also shows that planets must have highly compact solid cores with radioactive elements and hydrocarbons distributed throughout (to be discussed in detail).

CONCLUSION TO PART I

Concerning the traditional astrophysical theories (the nebular collapse theory of OSS and the 4.5 billion-year-old age of the planets, the density wave concept of galactic evolution, the Big Bang, the ice ball comet model, and the greenhouse effect) this paper claims that none of these are valid in spite of decades of theoretical effort. Two historical developments are responsible for this: 1) these theories were developed and given "accepted" status *before* the past decade's exploration of the solar system and 2) after excellent close range data were secured from space probes, the data were inevitably forced to fit the "accepted" theories by the advancement of hundreds of after-the-fact ad-hoc explanations (the oldest being the greenhouse effect for Venus' high temperature). It was far easier to do this than to ponder the failure of traditional theory. If any scientists raised objections to this, they were quickly removed from the astronomical community and life went on as before.

This paper shows that comets are not ice balls melted by solar radiation, but are asteroidal bodies which become electrically charged within the Sun's sphere of influence and are *attracting* the dust and ions observed in the comet tail. (The nebular ion cloud which lies past the orbit of Pluto is one source of comet tail material.) This causes a *build up* of material on the asteroidal comet nucleus.

Parts II and III of this paper will show that comets eventually evolve into planets (Venus may well be only a few thousand years old), moons, and asteroids, and that the solar system is dynamic, undergoing radical changes when large comets are captured into the inner solar system.

. . . to be continued.

REFERENCES AND NOTES

1. J. M. McCanney, "Continuing Galactic Formation," *Astrophys. Space Sci.*, 74 (1981), pp. 57-64.
2. J. M. McCanney, "Saturn's Sweeper Moons Predicted," *The Moon*, 24 (1981), pp. 349-53.
3. J. M. McCanney, "Electrical Phenomena at Saturn," unpublished.
4. When this paper was written in 1981, the common belief among astronomers was that the intergalactic boundary of the Sun lay somewhere beyond Jupiter. Pioneer 11 has passed the orbit of Pluto but to date has not encountered the heliopause. Two factors affect the location of this boundary: 1) the Sun's magnetic field interacting with the intergalactic wind (commonly called the "bow shock"), and 2) the force of the solar wind holding back dust and gas particles from entering the inner solar system. [Cf. "The Sun's Magnetic Field", KRONOS II:3 (1977), pp. 78-80. – *LMG*]
5. H. C. Houben, *Tidal Dissipation in the Solar System and the Possibility of Tidally Driven Planetary Magnetic Dynamos*, Cornell University thesis (Ithaca, 1978).
6. T. J. J. See, *Researches on the Evolution of the Stellar Systems*, Vol. II (Lynn, Mass., 1910), pp. 274-92.
7. R. A. Lyttleton, *Moon and the Planets*, 18 (1980), p. 13.
8. B. R. De, *Astronom. Soc. of India*, 3 (1975). (It is observed that De calls for the requirement of non-gravitational forces in any pseudo-"Laplacian" explanation of OSS. The traditionally accepted concept of the nebular collapse theory of OSS is based on Laplace's 18th century idea of a whirling cloud collapsing into a revolving solar system. In 1910 – based on the theoretical work by Newton and Darwin (no relation to Isaac; son of Charles) – T. J. J. See showed the classic flaws of the Laplacian theory, the primary difficulty being due to angular momentum considerations. He also proposed the first capture theory of OSS. See's work was ostracized and never promoted. All subsequent theoretical efforts have been attempts to account for the unexplainable in the Laplacian scenario. There is a great deal of speculation in today's "accepted" theory.)
9. B. A. Smith, *Science*, 204 (1979), p. 969.
10. In 1983 the Infrared Astronomy Satellite (IRAS) was launched and has since mapped infrared heat sources much more accurately and with greater sensitivity than could be done previously from mountain top infrared observatories. Almost immediately, wide publicity was generated stating that the birth of stars had been recorded. This is a great overstatement of the facts. It is true that new infrared sources have been recorded near magellanic clouds where traditional astrophysical theory would indicate and where none had previously been observed from mountain top observatories. However, to confirm the nebular collapse theory of stellar formation, two requirements are needed: 1) For IRAS to observe a new infrared source that had not previously been observed *by IRAS*

within its experimental limits of accuracy, and 2) that it was in fact the collapse of gas clouds that formed the star. The second condition would be difficult to prove using IRAS alone.

It is possible, for example, that massive nova remnants could be highly instrumental in the secondary formation of stars in magellanic clouds and that stellar collapse may not be possible without such a nucleus. Great caution must be used in interpreting data and drawing far reaching conclusions.

11. R. A. Kerr, *Science,* 207 (1980), p. 292.

12. J. H. Hoffman, *et al., Science,* 203 (1979), p. 800.

13. R. Berry, *Astronomy,* 9, No. 2 (1981), p. 19.

14. As recently as 1981, greenhouse theorists claimed that Earth and Venus evolved differently since the greenhouse effect was viable on Venus (due to its proximity to the Sun) but that Earth was not affected (in spite of its early CO_2 atmosphere) due to its greater distance from the Sun. Then the elevated temperature of Saturn's moon Titan was discovered. For lack of any explanation, the greenhouse effect was adopted to allow the data to conform to the 4.5 billion-year-age of the solar system which is basic to "accepted" theory.

15. D. Mulholland, *Science 82* (Dec., 1982), pp. 64-68.

16. A basic physics problem is to determine the length of time a planet of a certain size and mass can maintain an atmosphere. Both Titan and Pluto are well below the threshold that would allow a permanent atmosphere.

17. B. R. Sandel, *et al., Science,* 206 (1979), pp. 962-966.

18. R. Gore, *Nat. Geographic,* 157, No. 1 (1980), p. 10.

19. The magnetic fields of Jupiter and Saturn newly discovered by Pioneer 11 have created an unmentioned dilemma in the astrophysics community. In short, magnetic fields do not self-generate and sustain themselves for billions of years. Maxwell's equations have been swept under the theoretical rug by traditional theorists who still maintain that gravity is the only force acting in the cosmos.

    For magnetic fields to form, current must flow. For current to flow, a potential difference must be maintained. Herein lies the downfall of traditional astrophysical theory which does not allow any form of electro-magnetic interaction.

    When I pointed this out to a theorist in the physics department at Cornell University, he exclaimed that since both positively and negatively charged particles were moving in the recorded electric current between Io and Jupiter, no potential difference was required. This shows that too many scientists will say anything, no matter how absurd, to uphold traditional theory. Irrational behavior of this sort is found to be commonplace in astrophysics circles.

20. "Jupiter Pictorial," *Astronomy* (Aug., 1979), p. 54.

21. S. J. Peale, *Science,* 203 (1979), p. 892.

22. F. Hoyle, *Highlights in Astronomy* (San Francisco, 1975), pp. 39-49.

23. "Disappearing Mountains," *OMNI* (Jan., 1980), p. 38.

24. W. Ley, *Watchers of the Skies* (N. Y., 1966), pp. 334-8.

25. Venus Orbiter data have shed light on the interaction of solar radiation with the atmosphere of Venus. The result is that, although Venus has no protective magnetic field, no "coma-like" activity is observed. An unspoken dilemma among ice ball theorists is that neither the solar wind nor electromagnetic radiation is sufficient to create comas of the size observed in comets.

26. Z. Sekanina, *Icarus,* 37 (1979), p. 420.

27. F. Whipple, *Scientific American,* 242, No 3 (1980), p. 124.

28. Some IBCM theorists claim that wandering is due to ejection of highly volatile gases that exist in packets within a dirty ball of water ice. This, however, contradicts the claim of other theorists that all highly volatile frozen gases in the ice ball are routed out by solar radiation beyond the orbit of Jupiter and cause comets such as Kohoutek (1973) and Bowell (1981) to be observable beyond Jupiter and Saturn.

29. P. Koutchny, *Astron. and Astrophys.*, 72 (1979), p. 45.

30. P. L. Lamy, *Astron. and Astrophys.*, 72 (1979), p. 50.

31. R. A. Lyttleton, *Mysteries of the Solar System* (Oxford, 1968), p. 116. This book remains an excellent source of food for thought.

32. Middlehurst and Kuiper, *The Moon, Meteorites and Comets* (Chicago, 1963), p. 602, plate 4.

33. P. Moore, *Comets* (N. Y., 1976), p. 96.

34. J. E. Bortle, *Sky and Telescope*, 61, No. 3 (1981), p. 212.

35. G. Smith, *Nat. Geographic*, 157, No. 1 (1980), pp. 98-99.

36. W. H. Ip, *Planetary and Space Sci.*, 27 (1979), p. 121.

37. L. C. Lee, *Astrophys. J.*, 228, No. 1 (1979), p. 935.

38. The October, 1981, article by D. A. Mendis, *et al.* *(Astrophysical Journal*, 249, p. 787) on the charging of the comet nucleus, is quite different from the present paper. That paper assumes the ice ball comet model and claims charging to be a result of the solar wind and UV light impinging on the ice ball. It does not anticipate effects observable from Earth and does not allow the existence of the nebular ion cloud which is proposed in this paper. In brief, there is no apparent common ground between the two papers. It is curious to note, however, that the *Astrophysical Journal* published that paper but refused even to comment on the present paper which takes exception to the ice ball comet model.

39. One of the prominent astronomical observations of 1982 related quasars to normal galaxy evolution, in which young stars were forming in a cloud at the visible surface of certain quasars (ref. Boroson and Oke of California Institute of Technology). This is radically different from the density wave model's expectation that clouds of galactic dimensions should encounter another galaxy to force the formation of the density wave. It also supports Dr. H. Arp's observations that quasars are *not* objects at the "edge of the universe" (as proposed by those who use the Red Shift and Hubble constant as a measure of astronomical distance). Arp has observed "strings" which visibly connect quasars to galaxies, indicating that the Red-Shift-Hubble concept (*i.e.*, the Big Bang) is erroneous. He was threatened with loss of his telescope time by astronomers who disagreed. (Section IV of this paper discusses this in light of the newly proposed induced electric dipole red shift. For further reading, see *N. Y. Times*, section C (Oct. 19, 1979) and March, 1983 issue of *Sky and Telescope*.)

40. This is supported by the fact that only Jupiter spins about an axis parallel to the axis of the Sun.

41. Concerning the paper "Saturn's Sweeper Moons Predicted", it is now apparent from Voyager I and II data that the moons were not verified as predicted; however, the general concept of fusion in Saturn's atmosphere ignited by lightning (and the predicted side effects such as the rotating spokes) has been verified. Unfortunately, NASA scientists continue to maintain that all observed effects are "magnetic" in nature. Once again they erroneously assume that the magnetic fields simply exist, and that these in turn cause the observed electric effects (*i.e.*, the alteration of charged particle counts near Saturn's small moons, the electrical discharge that occurs in Saturn's atmosphere every time Dione passes overhead, the current sheet flowing constantly between Dione and Saturn – announced in October, 1982, and similar to the Io-Jupiter current sheet – the giga-amp electrical discharges that snap regularly among Saturn's rings, the correlation between electrical discharges and the "rotating spokes", the non-Keplerian orbits of certain ring particles and small moons, and the twisted F-ring). To date, NASA has not offered any viable explanations for these observations, although some were discovered over two years ago.

42. W. Filius, *et al.*, *Science*, 207 (1980), p. 428.

43. S. E. DeForst, *J. Geophys. Res.*, 77 (1972), p. 651.

44. In September, 1980, this was pointed out to the staff of *ICARUS*, which is edited at Cornell University. Their reply was that they did not know of, or believe in, such an effect on satellites. (They were not familiar with the DeForst paper or the fact that at least some members of the JPL imaging team observed charged particle

measurements to be affected by an unidentified electric potential at the surface of the space craft studying Saturn as referenced by Filius, *et al.*) In 1982, during the first space walk by space shuttle astronauts. Observations were made and photos were taken of an aura-like glow at the surface of the shuttle. NASA space scientists were asked to explain this and conjectured that it may be related to oxygen ions impinging on the surface (implying a net charge on the shuttle, although charging was never specifically mentioned).

45. Similar to the discharge of a backyard electric bug killer which discharges when a bug enters the area between anode and cathode.

46. The capacitor that forms with the Sun in its center and nebular ion cloud surrounding it (past the orbit of Pluto) will have surfaces of electrical equipotential between cathode and anode. It is now known that the second charging process depends on the comet nucleus crossing the surfaces of equipotential. As the comet nucleus crosses the equipotential surfaces, it must continually adopt the potential of space that it enters.

    This also aids in the electrical breakdown of the capacitor, initiating the discharge between Sun and nebular ion cloud which is observable as the comet tail and, occasionally, the sunward spike.

    It is explained later in this paper that circularization of orbit is a by-product of the drag the comet tail exerts on the comet nucleus. As the comet achieves a circular orbit, it no longer will cross the surfaces of equipotential and will remain essentially at a single potential. This is a simplistic explanation but applies well in the inner solar system where the equipotential surfaces are nearly spherical (past Jupiter, the Sun's magnetic tail will distort these surfaces). [Cf. Chris S. Sherrerd, "The Electromagnetic Circularization of Planetary Orbits", KRONOS IV:4 (1979), pp. 55-58; Ragnar Forshufvud, "On the Circularization of the Orbit of Venus", KRONOS VII:2 (1982), pp. 3-28. – *LMG*]

47. Past the orbit of Pluto, known from Pioneer 11 data.

48. It is now apparent that the observation of the higher proton current in the solar wind should have been interpreted properly a long time ago; but, as with many other cases of interpretation of data, scientists make data fit their theories and not vice versa. It has been known for a long time that the velocities of protons in the solar wind are much greater than those of electrons; however, scientists have maintained equal currents of electrons and protons because of the a priori assumption that space is electrically neutral.

    Since current = charge x velocity ÷ *l*, there is an excess *current* of protons in the solar wind, leaving the Sun with a net negative charge. The separation of charge must be an essential aspect of fusion in stars.

    The Sun cannot continuously expel an excess current of protons indefinitely, so there must be a current sheet of charge that flows continually between the nebular ion cloud and the Sun. In 1982, Pioneer 11 data confirmed the existence of such a current sheet flowing lengthwise along the Sun's magnetic tail. This may possibly relate this paper to the work of Juergens (KRONOS VIII:1) who assumed that such a current was necessary for his tufted anode concept (although I disagree that this is the source of the Sun's energy).

    The existence of the solar corona has always posed a problem to traditional astrophysics. This pure electron cloud with a temperature of millions of degrees is certainly not held in place by gravity and there is no apparent containment mechanism from without. The corona may act as a filter for the solar wind, accelerating protons while retarding electrons. This would account for the higher velocities of protons observed in the solar wind. At any rate, what holds the corona in place remains an unsolved mystery. [However, evidence is accumulating that the Sun's magnetic field plays the fundamental role in heating and containing the corona. See *The Sciences* (Dec. 1981), pp. 15-18, 32; *Natural History* (Jan. 1983), pp. 74-79; and *Scientific American* (Feb. 1983), pp. 104-119. – *CLE*]

49. The combined $1/r^2$ electric fields of the Sun and comet nucleus define the characteristic shape around the comet nucleus as designated in *Figure 1.*

50. R. A. Lyttleton, *Mysteries*, p. 113. 51. Middlehurst, *The Moon, Meteorites*, p. 602, plate 1C.

52. Past the orbit of Pluto (not Jupiter). 53. A. Dauvillier, *Cosmic Dust* (N. Y., 1964), pp. 56-7.

54. M. K. Wallis, *Nature*, 286 (1980), p. 207.

APPENDIX I

Proof No. 1

For all points on the circle of equal force, the resultant acceleration due to similar charges q_1 and q_2 is always in the direction of the Lagrange point (*Figures A1-1* and *A1-2*).

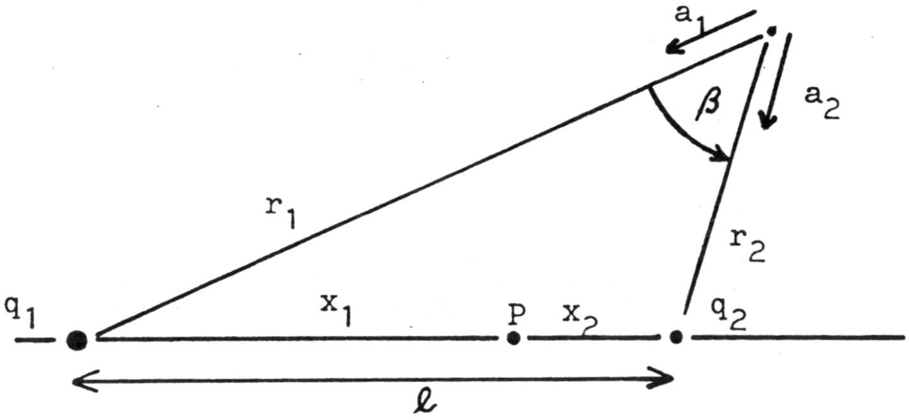

Figure A1-1

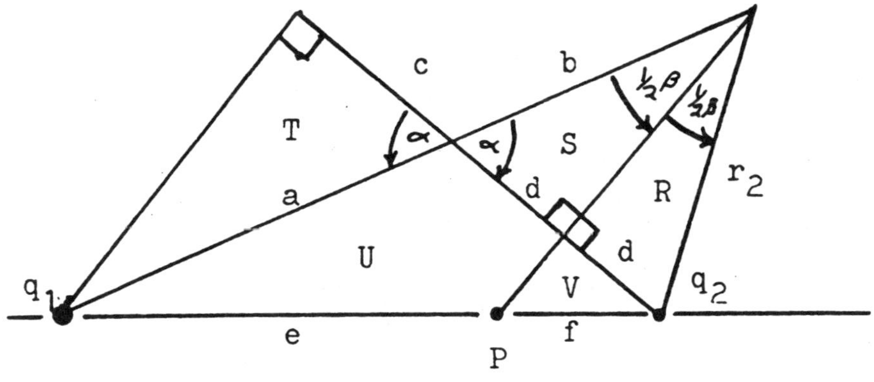

Figure A1-2

From Coulomb's Law:

$$a_1 = \frac{G\,q_1}{r_1^2}\ ,\quad a_2 = \frac{G\,q_2}{r_2^2} = \frac{G\,q_1\,(\frac{1}{K})}{r_2^2}$$

It is seen that $a_1 = a_2$ when $r_2 = \sqrt{\frac{1}{K}}\, r_1$

Thus $x_1 = \sqrt{k}\, x_2$ also.

Also, when $a_1 = a_2$, the vector sum of $\bar{a}_1 + \bar{a}_2$ will be the bisector of the angle β.

To prove that $\bar{a}_1 + \bar{a}_2$ points towards the Lagrange point for all points that satisfy the relation $|\bar{a}_1| = |\bar{a}_2|$; construct the following equilateral triangle (*Figure A1-2*).

Given: The two triangles R and S form an equilateral triangle, the construction is made such that triangle T contains one right angle as shown

$$a + b = r_1$$

$$r_2 = \frac{1}{\sqrt{k}}\, r_1$$

$$b = r_2$$

It is obvious from the figure that triangles T and S are similar.
Thus

$$\frac{a}{b} = \frac{c}{d} \qquad \text{but} \qquad \frac{a}{b} = \frac{(1 - \sqrt{\frac{1}{k}})\, r_1}{\sqrt{\frac{1}{k}}\, r_1} = \sqrt{k} - 1 = \frac{c}{d}$$

Therefore,

$$c = (\sqrt{k} - 1)d.$$

Also from the figure triangles U and V are similar. Thus

$$\frac{c+d}{e} = \frac{d}{f}\ , \qquad \text{or} \qquad \frac{c+d}{d} = \frac{e}{f}$$

$$c + d = (\sqrt{k} - 1)d + d = \sqrt{k}\, d$$

Therefore,

$$\frac{e}{f} = \sqrt{k} \qquad \text{or} \qquad e = \sqrt{k}\, f\ .$$

* * *

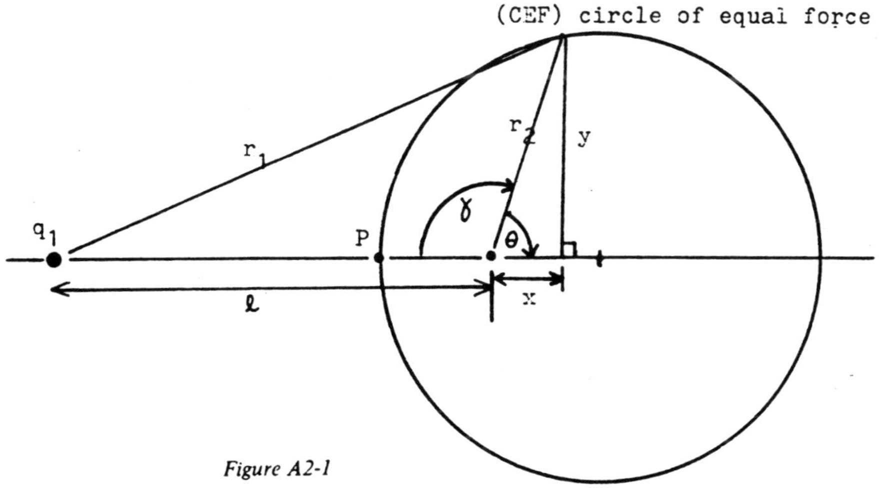

Figure A2-1

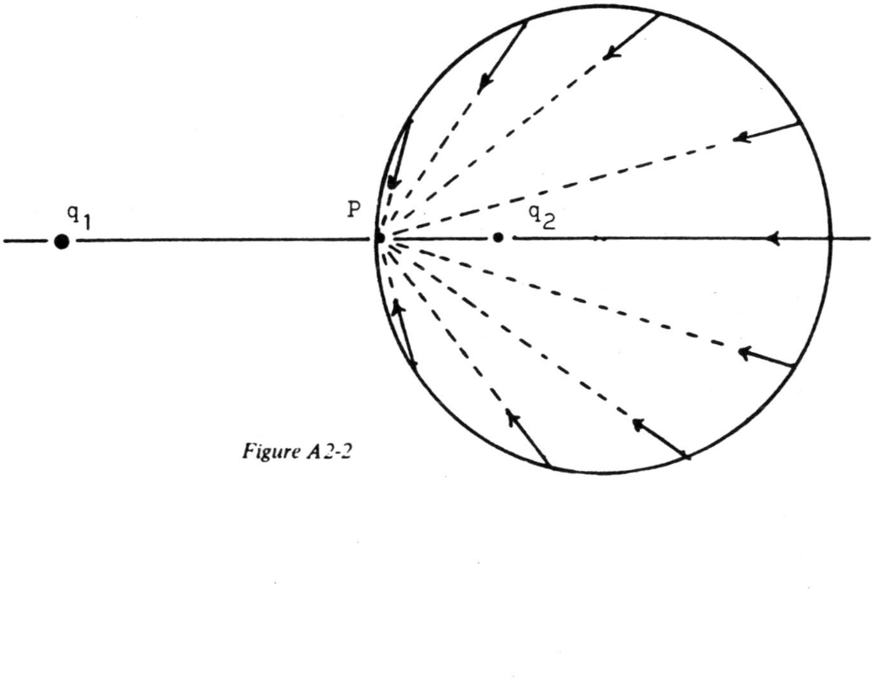

Figure A2-2

Proof No. 2

Equations 1-5 in the text, which refer to *Figure 2* of the text, are derived using Proof No. 2. Using *Figure A2-1:*

1) $r_1^2 = \ell^2 + r_2^2 - 2\ell r_2 \cos \gamma$ law of cosines

2) $kr_2^2 = \ell^2 + r_2^2 - 2\ell r_2 \cos \gamma$ ($r_1 = \sqrt{k}\, r_2$ from before)

3) $$\cos \gamma = \frac{\ell^2 + r_2^2 - kr_2^2}{2\ell r_2} = -\cos \theta$$

4) $$\frac{\ell^2 + (1 - k)r_2^2}{2\ell r_2} = 1/2\ (T + (1 - k)\ \frac{1}{T}) = -\cos \theta\ ,$$

(where $T = \frac{\ell}{r_2}$)

5) $$x = r_2 \cos \theta = r_2(-\ 1/2\ (T + (1 - k)\ \frac{1}{T}))$$

$$= (-\ 1/2)(\ell + (1 - k)\ \frac{r_2^2}{\ell})$$

6) Using $x^2 + y^2 = r_2^2$,

$$x = (-\ 1/2)(\ell + (1 - k)\ \frac{(x^2 + y^2)}{\ell}$$

7) $$1/2\ \ell + \frac{x^2 + y^2}{2}\ (\frac{1 - k}{\ell}) + x = 0.$$

8) Adding $(\frac{\ell}{k-1})$ to each x term in the equation will give the above equation in terms of the point $(\frac{\ell}{k-1}\ ,\ 0)$ giving the equation of a circle:

$$x^2 + y^2 = \frac{\ell^2 k}{(1-k)^2}$$

centered on the point $(\frac{\ell}{k-1}\ ,\ 0)$.

The equations of the text are easily derived from this equation.

THE NATURE AND ORIGIN OF COMETS AND THE EVOLUTION OF CELESTIAL BODIES

(PART II)

J. M. McCanney

PREFACE TO PART II

Part I (KRONOS IX:1, Fall 1983, pp. 17-39) presented critiques of the ice ball comet model (IBCM) and nebular collapse theory of the origin of the solar system (OSS) and argued that these "accepted" theories fall short of explaining numerous observed phenomena. Part I also introduced a new theory for comet behavior and solar system evolution based on the capture of comets. Comets were postulated to be discharges of a solar capacitor, the capacitor forming with the negatively charged Sun surrounded by a doughnut shaped nebular cloud of ionized dust and gases lying past the orbit of Pluto. Cometary discharges could also occur between the Sun and ionized matter of the zodiacal disc which rings the Sun. The major theoretical result was that charged comet nuclei are *attracting* the dust and gases in the comet tail and are not melting away as proposed in the IBCM. That is, comets are evolving into the planets, moons, and asteroids of our solar system.

Part I further extended concepts introduced in an earlier paper[1] that Saturn and its ring system (and Jupiter to a lesser extent) exhibit star-like properties including electrical discharges between its rings and in its atmosphere. The star-like properties are necessarily a result of localized fusion reactions in the gas planets' atmospheres which are ignited by energetic lightning bolts (which have been detected by Voyagers I and II). The analogy was made between the Saturn-ring and Sun-zodiacal disc systems. Part I should be read with its footnotes and references to put Parts II and III (in press) into proper perspective.

Part II further develops the new comet capture theory for the origin of the solar system (OSS) and proposes mechanisms for observed phenomena which must be accounted for in any self consistent theory. *Appendix I* provides a sample calculation, showing that the "tail drag" on a comet can explain two phenomena: 1) the Oort effect and 2) the rapid orbital circularization of comets with extensive tails. *Appendix II* details the experimental results of upcoming comet probes to Halley's Comet which can confirm this theory.

III. EXPLANATIONS OF OBSERVED PHENOMENA

Any alternate theory concerning comet behavior and the origin of the solar system (OSS) must reexplain many observed phenomena in a self-consistent context. These include the origin of comet nuclei and the reason for the observed "families" of comets arriving from many specific directions in space, comet wandering, sunward spikes, sunward fan tails, occasional separation of the tail from the nucleus, comet splitting, the cause of Type I, II, and III tails, the spiralling of tail material, the stratification in some tails, multiple tails, the shrinking of the coma as the comet approaches the Sun, and the maintenance of meteoroid streams.

In relating the above to the formation of planets, moons and asteroids, the theory must also explain the internal heat and radioactivity of the planets, the orientation of the rotational axes of the planets, the spacing of planetary and lunar orbits, the asteroid belt, the source of planetary atmospheres, the size distribution of celestial bodies, the cause of retrograde orbits of selected moons, and last, but not least, the magnetic fields of the planets. This must all be done in a context consistent with data (although not necessarily with uniformitarian theory) in other fields such as geology, biology, archaeology, anthropology, etc.

III a) *Sources of Comet Nuclei*

One source of comet nuclei has been identified as the dispersive spray of small conglomerates being ejected from the newly forming twin star systems at the base of the galactic arm.[(2)] Another possible source, which explains the existence of families of comets arriving from the same directions in space,[(3)] is the stellar nova. This is also the logical place to look for asteroidal type bodies which are ejected into interstellar space at high velocity.

The detectable remnants of a nova (identified as numerous point radio sources around the central nova star) indicate that large pieces of the solid stellar core remain in the vicinity of the explosion.[(4)] This strongly suggests that there must be many smaller fragments also. Two results of the new galaxy concept[(2)] are that the stellar core is not one of collapsed hydrogen, but is a solid planetary type core since all the celestial bodies are initially formed in the same way, and that the heavy isotopes – detected spectroscopically after a nova – come from this core; they are not generated in the explosion as previously thought.

From momentum considerations comes the result: the smaller the fragment, the greater its ejection velocity with the largest pieces remaining in the vicinity of the explosion. From analysis of a solid core exploding into a random assortment of pieces, a second result shows that there will be great numbers of small fragments, with fewer fragments of increasingly larger size. The nova will spray neighboring twin star systems with these fragments which may then be captured to begin new lives as comet nuclei.[5] Once again, it is a game of numbers. As with biological reproduction, of the multitude of seeds scattered by a plant, only a few will grow. The family groups of long period comets coming from the same direction in space and arriving in relatively closely packed groups,[3] therefore, are coming from the site of a nova which occurred millions of years ago.

A major result is that all the celestial bodies (stars, gas planets, terrestrial planets, moons, asteroids, and comets) are catalogued in one common grouping which is conceptually satisfying (as opposed to the numerous special cases defined in the nebular theory of OSS). So the capture of a large comet nucleus by a solar system is possible but less likely than the capture of the small nuclei.

Our solar system bears this out with over 5,000 asteroids, 1,000 comets, 50 moons (of which only about 10 are large), 9 planets (of which only 4 are large), and one star. Even within the asteroid belt, studies have shown that the number of asteroids increases as a geometric progression with decreasing size.[6] For this reason, the few close encounters of comets with planets that have been observed in the past 300 years show only comets with small masses. This is a very short time astronomically and cannot be extrapolated to the age of the solar system or to all comets as has been done by astronomers. If large comet nuclei pass through our solar system, a single passage with no close planetary encounters would cause no observable perturbation in the orbits of the present members of the solar system and, therefore, their masses would remain undetected.[7] It should be expected that the Jupiter-Galileo orbiter (presently scheduled for launch in the spring of 1986) will see comet nuclei (*e.g.*, small asteroidal bodies) as they enter Jupiter's radiation belts and detect the electromagnetic effects that must accompany such an encounter.

III b) *Explanation of Observed Cometary Phenomena*

There are three soūrces of planetary atmospheres. The primary source is the nebular ion cloud (explained in Part I, KRONOS IX:1) which contains the light elements up to approximately sulfur, as this is what is observed in comet tails of extreme length. That is, as the comet transforms into a hot young planet with a circularized orbit (to be discussed in detail), the original planetary atmosphere will already be present (Venus being a prime example).

Another source is the chance grazing of the Sun, as in the case of the comets of 1882 and 1887; some grazing comets are actually seen to fluoresce after leaving the Sun, indicating the possible presence of a newly acquired atmosphere. Isaac Newton in his *Principia* includes a description of the Great Comet of 1680 as it plunged through the solar atmosphere. Ironically, in the next sentence he describes a nova explosion which, unknown to him, will provide comet nuclei for future solar systems.[8]

It is generally claimed in the nebular theory that the atmospheres of Earth and Venus were released from their cooling interiors during volcanic activity. Gases are certainly released by volcanic activity even today, but this article contends that much of the original atmosphere was formed as part of the comet phase of planetary evolution. If Venus joined the planets of our solar system only a few thousand years ago then it already had a huge CO_2 atmosphere amassed from its comet stage of development.[9]

Spiralling of tail material is sometimes observed in comets. Before 1950 a number of comet theories invoked magnetic fields[10] as the cause of spiralling. All assumed *a priori* that the tail material moved away from the comet nucleus. The politics of science, however, suppressed such notions of magnetic fields. This paper resurrects those ideas and provides a theoretical basis for them, but with one major difference: the tail material does not move away from, but is drawn towards the comet nucleus. The reader should note that Nobel Laureate Hannes Alfvén stands as one of the few investigators who has long recognized the influential nature of electro-magnetic fields in the phenomena observed in comet tails.

From just observing the spiralling motion, it is impossible to tell whether the cometary magnetic field is caused by circulating charge on the comet nucleus or the current flow of the Sun's capacitor discharge, since the resultant magnetic force will be central in both cases for electrons in the spike and ions in the tail. Here the reader is referred to the American Indian rock paintings (*Figure 1*) and a similar drawing of the Great Comet of 1861 (*Figure 2*) and the spiked

Figure 1

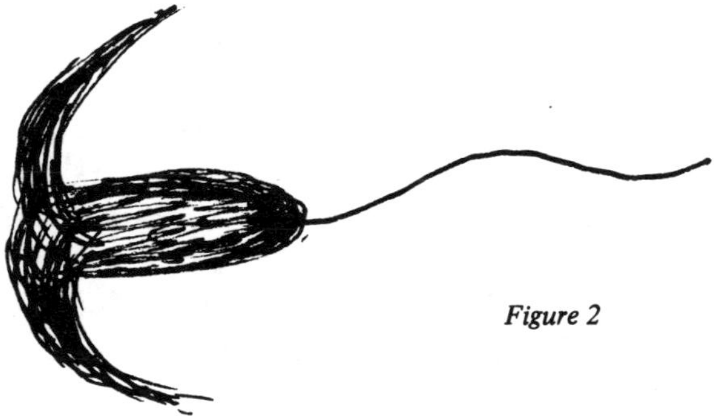

Figure 2

Figure 3

Figures 1, 2, and *3* above illustrate comets with sunward spikes and plumes. *Figure 1* is a prehistoric American Indian rock painting found near Greenriver, Utah (photo by Gary Smith, Windsinger Enterprises, Inc.). Note the snake-like sunward spikes. *Figure 2* shows the Great Comet of 1861 with its snake-like spike. *Figure 3* is Comet Arend-Roland (1957) with its intense sunward spike. Iceball comet theorists claim that sunward spikes are only optical illusions.

Comet Arend-Roland (*Figure 3*). These comets are not easily explained in the ice ball comet model (IBCM). (The topic of spiked comets will be discussed again.)

Comet wandering was first noticed in Encke's comet in the 19th century, and it was Encke's original idea to account for this by the "secular action of a resisting medium". George H. Darwin and T. J. J. See used this to develop the first capture theory of OSS (in the late 1800s)[8] showing two effects over very long time spans, on the order of millions of years for planet-sized objects: 1) elliptical orbits would be circularized and 2) circular orbits would slowly spiral sunwards. These are also known as eccentricity damping and energy disposal. A complete treatment of the celestial mechanics involved is also given by Smart.[11] The same effect occurs in the Poynting-Robertson effect for micron-sized particles acted on by solar radiation.

Encke's assumption was that the cross sectional area of the comet nucleus would drag through the resisting medium with the density increasing near the Sun. But this concept was incomplete because it is now known that the solar wind effectively clears the inner solar system of this resisting medium (except for the zodiacal disk and newly discovered rings between Mars and Jupiter[12]) and, even if it existed, it would have to be quite dense to cause the observed perturbations of cometary orbits.[8, 11, 12] The new comet theory shows the same effect, but with the tail drag being millions of times greater than Encke's original suggestion since vast quantities of matter are drawn into the comet nucleus from the rotating zodiacal disk and/or stationary nebular ion cloud which lies past the orbit of Pluto. So a comet with a large tail can have its orbit circularized or may seem to lose a great deal of energy in a very short time astronomically (*i.e.*, within 1,000 years). A sample calculation is provided in *Appendix I* to illustrate the effect of tail drag in circularizing the orbit of a comet as is suspected to have been the case with Venus only a few thousand years ago. Comet wandering is also caused by the variable tail drag as discussed previously. This is important in the development of the statistics governing orbital spacing and solar system formation (to be discussed).

The current theoretical efforts in the nebular theory of OSS explain the orientation of rotational axes of the planets and moons as an effect of localized magnetohydrodynamic phenomena early in the collapsing solar nebula.

The new comet theory explains this to be due to the non-symmetric vortex motion of infalling tail matter. Since the comet normally begins with a relatively small nucleus compared to the final planet or moon,[5] the final spin rate and orientation of the spin axis can take on random values which are governed by chance. Another factor that may cause the nucleus to rotate is an encounter with the solar atmosphere at high velocity during Sun grazing, but this must be rare compared to the primary cause of axial spin. Planetary precession is a very long term effect (to be discussed), but possibly tidal effects causing crustal shifting during comet capture can cause alterations in rotation rates and axis of spin.(13)

The maintenance of meteoroid streams is explained in the present paper by the *short range* induced electric dipole force which acts on electrical conductors (*i.e.*, meteoroids containing metals). The importance of this force has been discussed in an earlier paper(1) and was observed when Pioneer-Saturn passed a short distance under the small charged moon 1979-S2. At short range, this force becomes much greater than gravitational forces. So a highly charged comet nucleus will give a very strong impulse to meteoroids in its vicinity, allowing them to remain gravitationally bound to the Sun when farther away from the comet nucleus, with the meteoroids assuming the same orbit as the comet. As meteoroid size decreases, the mass decreases as the cube of the radius. Therefore, smaller objects are affected more by the induced dipole force.(1) Also, since the force is proportional to the square of the net charge on the comet nucleus, the forces on a meteoroid would be much greater than the impulse experienced by Pioneer when near Saturn's small moon. The net effect is the gathering and maintenance of meteoroid streams which will then follow the same orbit as the comet. If, at a later date, the comet nucleus is perturbed into a new orbit, the meteoroid stream will remain in the original orbit of the comet nucleus. This accounts for the numerous meteroid streams which are observed to orbit the Sun. Spacecraft design must take the induced electric dipole force into account, as mentioned previously, when these craft are expected to approach highly charged celestial objects.

The Earth's ionosphere will act as a conducting sphere in the non-uniform electric field of the Sun, the charged Moon, or a nearby comet nucleus. Ionospheric charge will adjust constantly to maintain an electric field of zero to the inside. What effect this has on weather or other phenomena should be investigated, since a well known fact of meteorology is that "jet streams" in the upper atmosphere move the surface level weather systems. This implies that what are normally termed *magnetic storms* in the Sun have electrical counterparts which escape detection (other than in disruption of radio broadcasts) for this reason.(14)

It appears that there are two causes of Type II comet tails, although data are sparse. These smooth curved tails, *e.g.*, Donati's comet, result from orbiting particles in the Sun's zodiacal disk being drawn into the comet nucleus.(15) If this is the case, then the curved tail may either lead or follow the comet depending on the relative angular velocity (prograde or retrograde) of the comet and zodiacal disk. A second type, such as seen in Comet West and Halley's Comet (1910),(15) shows striated structure which results from the movement of the comet nucleus and is most pronounced in comets near the Sun. It has been shown that this observed structure cannot be explained by gravitational or solar radiation pressure effects.(16) The present theory explains this as a result of the comet nucleus continually moving out from between the forming tail and the Sun, the tail material being first drawn inwards by strong electrical forces, then dispersed by the solar wind as the comet nucleus moves and is no longer properly aligned between the Sun and tail. Tail separation is also observed, occurring when the charge on the nucleus is suddenly neutralized, whereupon the tail, like the second Type II tail, is quickly dispersed by the solar wind.

IV. SOLAR SYSTEM FORMATION

IV a) *Comet Capture Processes and the Formation of Solar Systems*

Capture of long period comets by a twin star system is well documented and well understood. Energy transfers during close encounters with planets can either give energy to or take energy from the comet nucleus, so once a new comet nucleus is captured by the solar system, it will continue to encounter planets until it obtains a non-overlapping orbit or is expelled from the system. Lack of an exact solution to the N-body problem complicates the approach to understanding. Specific examples have been worked out,(17, 18, 19) but more work is needed to catalogue lunar capture routes.

One "calculated" result which I find hard to accept is that entire planets would be vaporized during close planetary encounters due to tidal friction. This has been used as an argument against lunar capture (especially retrograde).(20, 21) This paper argues that tidal friction has been greatly overestimated by investigators such as MacDonald, Wise, and by current theorists who claim that the great internal heat of Jupiter's moon Io is caused by tidal friction. (This will be discussed again.)

The radiation belts of the planets with magnetic fields play a fundamental role in planetary and lunar capture. Comets will be seen to brighten greatly when they pass through Jupiter's radiation belts, *e.g.*, Comet Brooks II (1889) underwent brightening as did Biela's Comet (1846) and the nucleus divided while in Jupiter's realm.[22, 23] The brightening is a result of the sudden charging and influx of tail material as the asteroidal comet nuclei encounter the Jovian radiation belts at high velocity, implying that velocity relative to the plasma is a factor in the charging process. With the charging process begun, the nucleus will be assured of inducing the discharge of the solar capacitor as it enters the inner solar system and therefore assures the proper development of the comet.[24] This same process helps slow comet nuclei during lunar capture and thus gives planets with large magnetic fields an advantage in lunar capture. It is apparent that, as the forming moon or planet achieves a circular orbit, it becomes "immune" to the charging process. What terminates the charging process is discussed in footnote No. 46 of Part I of this paper (see KRONOS IX:1). The moons of Jupiter and Saturn which lie in circular orbits move in plasma rings [25] which evidently is related to this immunity.

The slightly elliptical orbit of Io seems to be in agreement with this as well. The 5 million amp current flowing from Io to Jupiter[26] indicates that Io maintains a potential difference with respect to Jupiter, but this is much less than what would be expected for a highly charged comet. The tail drag will continue to circularize the orbit, becoming less effective as the circular orbit is achieved and the charging reduced. Io shows that electrical discharges *do* occur between celestial bodies.[27]

The ring patterns found on Callisto had been explained in a previous paper[1] as due to the approach and impact of a small highly charged asteroidal body in the ferromagnetic dust that covers Callisto's surface. I now believe that it was the result of a large discharge between Callisto and a charged asteroidal body, and that the "crater" in the center of these rings was formed as the discharge burned the planetary surface. (See *Figures 4, 5,* and *6.*) If these rings were *mechanical wave patterns* on the lunar surface caused by an impacting object when Callisto was young and molten as suggested by NASA,[28] then many superimposed wave patterns should have occurred due to other nearby impact craters (there is no lack of impact basins on Callisto's surface). The nebular theory assumes that this cratering occurred early in the history of the solar system. If the rings resulted from the alignment of ferromagnetic dusts in the magnetic field of an electrical discharge (as proposed by

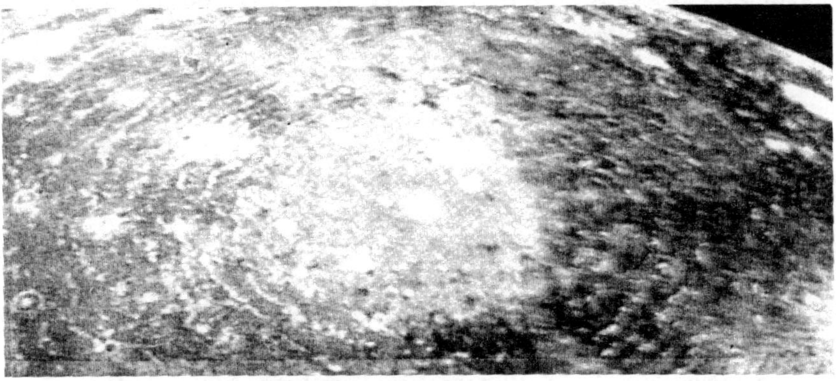

Figure 4

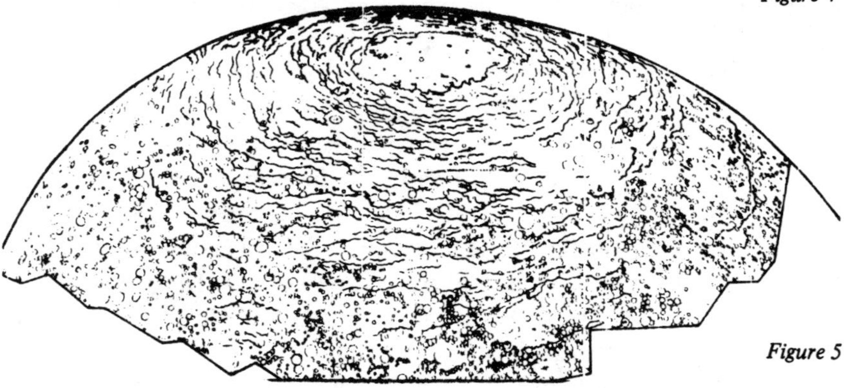

Figure 5

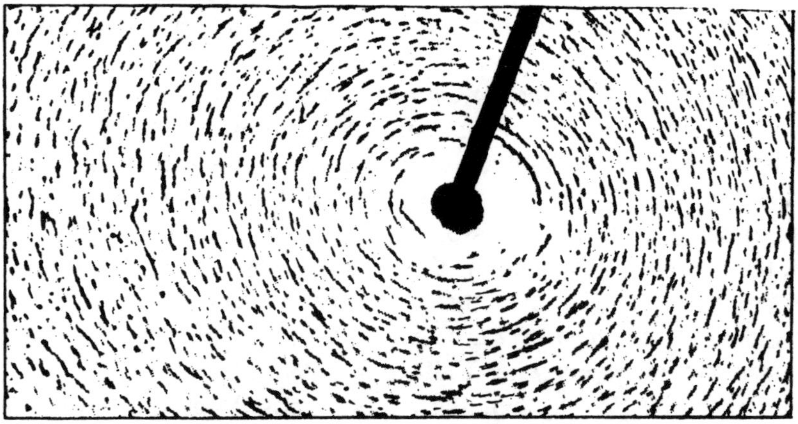

Figure 6

Figures 4, 5, and *6* above illustrate the "ring effect" caused by ferromagnetic dust in the presence of a strong electrical current. *Figure 4* is a photo by Voyager I of such a ring structure on Callisto. A similar but smaller pattern exists on Callisto's far side. *Figure 5* is a computer plot of the ring patterns on Callisto (courtesy of NASA). *Figure 6* shows a simple laboratory experiment, with iron filings forming rings around a current carrying wire.

the author), then all previous ring patterns in the vicinity would be destroyed,[29] leaving only one set of concentric rings, as is observed. This also gives support to the electrical discharge concept of the sunward spike of comets and the Venus to Earth discharges described by Velikovsky.

In the case of Callisto, if the passing comet nucleus was very large (as is rarely the case), then it would have also gravitationally perturbed the moons of Jupiter and these must have then re-circularized their orbits, since their new elliptical orbits would have allowed them to charge again within Jupiter's proton wind supported capacitor.

Explaining the retrograde motion of certain moons has been one of the great difficulties of the nebular theory of the origin of the solar system (OSS). Capture or encounter phenomena are usually assumed to account for these special cases. This, however, seems out of place in the theory whose main objective is to show the planets and moons evolving in already circularized orbits.[30] In the comet capture theory of OSS, retrograde motion is possible, but its chances of survival are small. The direction of spin of the solar system is governed by Jupiter. (Comets which evolve into moons with retrograde motion may gain energy during planetary encounters and, therefore, have a higher probability of being ejected from the system.) In summary, a moon can be captured by 1) energy loss due to sudden charging and mass accretion (*i.e.*, tail drag) as it enters a planet's radiation belts, or 2) energy loss associated with the gravitational encounter with the Sun-planet system, or 3) capture involving an energy transfer between the comet and a planet-moon system (similar to capture by a twin star system).

The asteroid belt is explained in the nebular theory of OSS as a region in which a planet never formed. According to the present theory, however, the asteroid belt is an area in which a planet could reside. Since none has been captured, it is available for asteroids to accumulate.

In the context of the present theory, the asteroid belt provides a remarkable statistical experiment, as it contains a random sampling of celestial bodies which have been captured continually since the beginning of the solar system. The importance of studying the asteroids, some of which exhibit halos, is apparent. The asteroidal orbits indicate statistically the possible orbits which result from capture processes by Jupiter and to a lesser extent by the other planets. The two asteroidal moons of Mars, and those of Jupiter, Saturn, etc. are undoubtedly captured wanderers of the asteroid belt. This also corresponds with the observation of the groups of comets associated with Jupiter,

Saturn, Uranus, and Neptune which are the results of many captures.

It is impossible to make a distinction between short period comet orbits and the orbits of many asteroids,[31] suggesting that the comet nuclei are indeed evolving into the asteroids. The Earth-Moon system is a likely candidate for capture of a new asteroidal moon and, if this occurred, would give an excellent chance to view lunar capture at close range.

Comet wandering is well documented and is explained in the IBCM as due to the ejection of vapor jets from the ice ball.[32] The present theory explains wandering to be the result of variable tail drag, the same effect which causes the circularization of orbit. The interaction of the charges of the Sun and comet nucleus may possibly be significant in highly charged comets with small nuclei, but Part I (KRONOS IX:1) shows by way of a calculation that electrical effects are noticeable only over very long periods of time compared to the dominant effect (*e.g.*, tail drag).[33]

IV b) *The Statistics of Solar System Formation*

The evolution of N-body systems has been studied extensively in celestial mechanics, using both mathematical derivations and computer simulations of orbits.[34] For systems like our own solar system in which the orbits lie more or less in one plane,[35] a stable system will evolve in which the larger bodies (the planets) are spaced apart from each other in non-overlapping orbits. The reason that the system evolves toward this type of stability is as follows: If two bodies have orbits that lie more or less in the same plane and their orbits are overlapping (or are in certain types of resonance), encounters or perturbations will occur between the two bodies, changing one or both orbits. If no two major bodies are in overlapping or resonant orbits, then no further encounters or secular perturbations will occur; the system is stable.

The rapid circularization of orbit caused by comet tail drag assures a rapid relaxation time for the system once a non-overlapping orbit is achieved by a newly captured member (*e.g.*, Venus). This is the cause of orbital spacing in planetary and lunar systems, the asteroid belt being a prime example of the possible orbits attainable by capture. Our solar system is at present quite stable and well ordered. Systems with very large twin stars in highly eccentric orbits are common and will form an interesting branch of study in the capture theory of OSS.

It has been noticed[36] that the Sun-Mars-asteroid belt-Jupiter system is spaced almost identically as the Saturn-Rhea-G ring-Titan system and that the masses are in a similar ratio. With the already stated analogy between the Sun and Saturn, it seems plausible that Saturn's G ring is a ring of small meteoroids captured by this system. Extrapolating the argument for the size distribution of the planets, there should be millions of small particles in the G ring. Oddly enough, T. J. J. See predicts this phenomenon[37] as he states: "Similar zones of asteroids . . . may be supposed to exist in other planetary systems having large planets such as Jupiter."

The Moon rocks gathered by Apollo astronauts have been dated at 4-4.6 billion years and this is interpreted by some as the age of the Earth-Moon system. The rock and dust on the Moon's surface, however, is just a random sampling of debris floating in space. The age of these rocks corresponds well with dates of meteorites found on Earth, as they are the same. With the enormous meteor streams that have been recorded on Earth,[38, 39, 40, 41] one cannot expect the original lunar surface to be showing, especially since there is no erosion or wind to move this debris. So to date the Moon, one would have to dig to sample the original lunar surface. The Moon rock data suggest, however, that 4.5 billion years is the average age of matter in the vicinity of the Sun and, therefore, is the best estimate of the age of the Sun-Jupiter pair. The oldest known Earth rocks are somewhat younger than this.[42, 43] Their ages give the true date for the Earth's transformation from a comet into a planet.

IV c) *Internal Energy Sources*

Venus, Earth, Jupiter, Saturn, and the Sun all have internal heat sources (if the data are interpreted literally), since all radiate more energy than they receive from external sources. As already mentioned, all four Pioneer-Venus probes measured more radiation rising from the planetary surface than entered as sunlight. The Earth's internal heat is known to be caused largely by radioactive decay, so it would not be presumptuous to assume that the same is true of some of the other planets, especially since they were all formed by the same process. Since the comet is the ultimate high energy accelerator,[44] an abundance of heavy radioisotopes is expected in the forming planetary interior which will guarantee an internal heat source for millions of years. The initial heat of formation must be very great and this must take some time to cool down. On such matters, future experimental measurements must provide the needed cooling data.[45]

Voyager I has shown that Saturn produces more internal heat for its size than does Jupiter (indicating its younger age as is expected from the comet capture theory) and has been shown to be a small star with ongoing fusion in its atmosphere.[1] In Saturn, both fusion (in the atmosphere) and radioactive decay of heavy isotopes (in the core)[4] occur which suggests that this must also be the case for the Sun. A major result of a previous paper[1] and subsequent Voyager data suggest that Saturn's, Jupiter's, and, therefore, the Sun's fusion is ignited and controlled by electrical discharges in their atmospheres. This conjecture must be true since both Jupiter and Saturn exhibit star-like characteristics[46] but have insufficient mass at present to support fusion deep in their cores by the commonly accepted mechanism.

Therefore, the state of a planet or star is seen to be not so much one of size, but age as related to size (as is apparent from bright comets that pass very near the Sun). A hot newly-formed planet may burn hydrogen for some time before it cools. The new planet passes through a super hot radioactive phase, followed by a chemical stage (Venus is presently in transition between these stages) allowing the free combination of elements, after which comes the cascade of biological evolution at temperatures below 200°F and evidently from a perfectly sterile medium. Thus, all planets begin with approximately the same chemicals in the same proportions and one should expect biological systems to reflect this.

Some scientists claim that the effects of tidal friction are heating Io, whose surface is volcanically active. As already mentioned, however, the volcanic activity occurs near the equator and not near the poles as predicted by the tidal theory. The six distinct theories of heating for the Sun, Jupiter, Saturn, Venus, Earth, Titan, and Io developed by nebular theorists reflect the *a priori* assumption that all these bodies formed 4.5 billion years ago, and therefore separate theories are needed for each. (These are: fusion in the Sun, primordial heat in Jupiter, belated helium collapse in Saturn, greenhouse effect for Venus and Titan, radioactive decay in Earth, and tidal heating for Io.) This is not to say that there are not six or more separate processes, but compared to the comet capture concept which develops a unified concept for internal heating, it is less satisfying.

An unmistakable high altitude wind pattern is visible on all the "hot" planets, with the wind belts slowly migrating from the equator to the poles.[47, 48, 49, 50, 51] Venus' belts migrate completely in a few days, whereas the migration of the Sun's belts takes 11 years. Also, Saturn's broad high velocity equatorial wind belt corresponds to the Sun's differential rotation (33 days at the poles and 25 days at the equator), showing that Saturn is more star-like than Jupiter which has less differential rotation.[46] The migrating wind belts must be the result of heat rising from the cores of these bodies. A complete simulation of circulation patterns caused by heat rising from the core of a planet containing excessive internal heat has been done by F. Besse[52] showing poleward migration of high altitude atmosphere and shears resulting in wind belts. (It has been suggested by some that the solar migration is due to "magnetic effects"; however, Venus has no magnetic field but exhibits similar properties.)

Venera and Pioneer detected lightning in Venus' atmosphere similar to that found in the atmospheres of the Sun, Jupiter, and Saturn, sometimes called "whistlers",[53, 54, 55, 56, 57] implying that these result from dynamic storm systems in a turbulent atmosphere. By the present analysis, Venus, Saturn, Io, Titan, and Pluto must all be relatively young bodies. Until alternate dating techniques are developed, exact ages cannot be given except possibly in the case of the planet Venus for which there is apparent historical information concerning its formation as a planet. All have internal heat sources in their cores (as does Earth) due to long term decay of heavy radioisotopes formed during the comet stage of development, and some exhibit high latent heat from recent formation. Also, the gas planets support fusion in their atmospheres. Tidal effects of Jupiter, Europa, and Ganymede do affect Io but this is not the major source of Io's heat. The gravitational tugging only assures constant agitation allowing the internal heat which is already present to escape rapidly to the surface by volcanism.

At numerous points this paper has referred to the formation of heavy radioisotopes in the comet nucleus from high energy particle bombardment during the discharge of the solar capacitor. This arises from the need to explain the known presence of radioactive elements in the Earth and Moon, and the realization that very high energy electrons, protons, and ions will be impinging on the comet nucleus in the highly *active* comet. (Low activity comets will only experience compaction and chemical bonding due to infalling particles; *e.g.*, chemical bonds form in energy ranges from 2 to 10 eV.)

Nuclear alterations begin to occur at particle energies as low as 200 keV (transmutation from ${}_5Be^{11}$ to ${}_4Be^8$); however, the majority of nuclear transformations occur in the energy range from 20 to 50 MeV (million electron volts). Internal reactions of atomic nuclei occur at energies as high as 7.5 MeV in the unstable very heavy isotopes such as Californium, Fermium, etc.;[58] however, it is the Coulomb barrier which must be overcome by particles bombarding the atomic nucleus. These generally range from 15 to 45 MeV for various atomic nuclei, requiring incident particle energies of 20 to 50 MeV to produce a wide variety of nuclear transmutations in the highly active comet.

A previous paper[1] noted that measurements have been made as early as 1970 showing that satellites can charge quickly to voltages as high as 10,000 volts[59] while moving in the plasma which surrounds Earth. That article[1] showed by way of a calculation that surface potentials in excess of 1.0 MeV were easily obtainable by small moons in Saturn's radiation belts, and that this was noticed by Pioneer 11 as it passed the moon 1979-S2. Voyagers I and II later confirmed the alteration of charged particle counts near all of Jupiter's and Saturn's moons. This charging was explained by the same mechanism used to create the observed spokes in Saturn's rings and the same charging mechanism used by comets.[1]

It is therefore expected that electrical charges and, therefore, particle energies in *highly active* comets will be found in excess of 20 MeV. This theory predicts that the particle energies found in highly active comet nuclei will far exceed the energies available in accelerators built for nuclear study on Earth, and may provide for free the ultimate particle accelerator. The abundance of heavy radioisotopes found in the interiors of Earth, the Moon, and expected to be found in other celestial bodies is a result of formation from a comet, which is a discharge of the solar capacitor.

Note that *Appendix II* lists the predicted results of experiments of the four Halley's Comet probes which are expected to confirm the expectations of the present theory of comet behavior and solar system evolution.

CONCLUSION TO PART II

Part II further developed the new comet theory to provide self-consistent explanations for: the sources of comet nuclei, observed cometary phenomena, the formation of solar systems by comet capture, the process by which comets evolve into planets with circular/non-overlapping orbits in astronomically short times, and the internal heating of the planets and moons of the solar system.

Part III will introduce the induced electric dipole red-shift concept[60] and show how it accounts for the anomalous red-shift data that currently are under controversy in the astronomical community. It will also discuss a theory for geomagnetic field formation with the dynamo powered by a planet spinning inside a slightly charged moon, and show a correlation between magnetic field reversals and the external effects of a highly charged comet passing close to the planet. It will further discuss biological evolution and celestial catastrophism, and the collective fear of Velikovsky that has permeated the scientific community for the past 35 years.

. . . to be concluded.

REFERENCES AND NOTES

1. J. M. McCanney, *The Moon and the Planets,* 24 (1981), pp. 349-53.
2. J. M. McCanney, "Continuing Galactic Formation," *Astrophys. Spa. Sci.,* 74 (1981), pp. 57-64.
3. R. A. Lyttleton, *Mysteries of the Solar System* (Oxford, 1981), pp. 135-38. (Numerous scientific papers have also appeared recently on this subject.)
4. In a footnote in a previous article by the author (see ref. No. 1 above, p. 350), it was stated that the Sun, Saturn, and Jupiter must have solid planetary cores as implied by the galaxy paper (ref. No. 2 above). It also predicted at that time that Saturn (and Jupiter to a lesser extent) would exhibit star-like properties similar to the Sun. Since then, Voyagers I and II data have convinced many other scientists of the same.

   The June 23, 1983 *New Scientist* (p. 856) reported that Carl Rouse arrived at the same conclusion as a result of theoretical work on the solar neutrino problem. This problem was discussed earlier by E. Milton in KRONOS V:1 ("The Not So Stable Sun"). The papers by the author suggest a different explanation for the "lack of solar neutrinos". The gravitational collapse theory of stellar formation suggests that neutrino production should be high because the bulk of thermonuclear reaction is said to take place deep in the Sun's interior where pressures are greatest. In this paper and previous papers, the author claims that fusion in the Sun (and also in Jupiter and Saturn) is ignited by highly energetic lightning bolts in their atmospheres. This would relax the requirement of fusion occurring deep in the stellar interior, and allow for the solid core and isodense atmosphere which would fit solar oscillation data (see Milton's paper).
5. From seismographic data of the Earth, it is well known that the density of Earth's interior is quite constant, but has a small, very dense central core. This core is the original "planetary seed" of Earth which was the original comet nucleus. It is interesting to note how large the Earth is compared to the original seed nucleus. As mentioned in Part I (KRONOS IX:1), comet development depends more on the discharge of the solar capacitor than on the size of the comet nucleus.
6. A. Dauvillier, *Cosmic Dust* (N.Y., 1964), pp. 26-7.
7. The scientific papers and popularized book by S. V. M. Clube and W. M. Napier [*Nature, 282* (29 November, 1979), p. 455; *Q. Jl. Astr. Soc.* (1982), *23,* pp. 45-66; and *The Cosmic Serpent* (London and New York, 1982)] deserve comment in light of the present paper. Their statement to the astronomical community that Velikovsky was at least "close to correct" and that comets (and in particular a body named Venus) did play important roles in early civilizations is long overdue. They also recognize the "irrational" behavior with which "scientists confronted Velikovsky", and the overwhelming data pointing towards worldwide catastrophism caused by comets.

   I maintain, however, that Clube and Napier's findings support the present paper more precisely than their own theory of terrestrial catastrophism. Both papers agree that comets are of interstellar origin, and that the primordial Oort cloud concept does not explain the observed comet flux in the solar system. They support the ice ball comet model, however, and modify the original Oort cloud concept to say that it is periodically replenished. This paper claims an entirely new concept for cometary phenomena and infers that comet nuclei are asteroidal bodies (not ice balls) that arrive at our solar system in closely packed groups from past novas of neighboring stars (also, occasionally, comet nuclei may reach planetary dimensions).

We agree that the data clearly show that new comets arrive sporadically, but there are underlying differences. Clube and Napier say that a period of approximately 50 million years is evident and caused by our solar system passing through galactic arms (their second article departs from the density wave model of galactic formation and suggests that other models need further consideration). Table 4 (page 458) of their *Nature* article lists the following ages of worldwide catastrophic events corresponding to mass extinctions (ages given in millions of years) = 1, 13, 25, 35, 58, 63, 135, 181, 230, 280, 345, and 405. Note that by taking the difference between successive events, a regular period of 50 million years is not obtained, but it suggests a *random* distribution (differences in millions of years) = 12, 12, 11, 22, 5, 72, 46, 49, 50, 65, 60. Only 3 of the 11 epochs are close to their 50 million year period. The present paper says that such events are governed strictly by chance. Their data bear this out dramatically.

Their work and this paper agree that short period comet orbits cannot be distinguished from orbits of the asteroid belt and Apollo (near Earth) asteroids. We both claim (although by different mechanisms) that comets are evolving into these asteroids. This paper goes on to claim that comets can also evolve into planets and moons (the Velikovsky connection).

Both papers agree that comets have been the cause of worldwide catastrophes and change. Clube and Napier depend on direct collisions only, whereas this paper allows that a few comet nuclei can become very large, and that close encounters can also cause major damage to Earth by gravitational effects and electrical discharge effects (in support of Velikovsky).

Finally, Clube and Napier do not support *a priori* the nebular collapse theory of OSS (p. 59 of *Astr. Soc.* article) as do most astronomers. We both agree that, today, at least one comet poses a potential threat to Earth and that history will repeat itself as surely as there are stars in the sky.

The solar system is a dynamic place and its true history is becoming clear. When Athens' great statesman Solon visited Egypt in 572 B.C. to inquire about their knowledge of the flood, the Egyptian historians told him of the Great Deluge, the destruction of Atlantis, and five different catastrophes which had plagued their development in the previous 10,000 years. There is a scientific basis now for believing that statement.

8. T. J. J. See, *Researches on the Evolution of the Stellar Systems,* Vol. II (Lynn, Mass., 1910). (Pages 134-5 discuss Isaac Newton's observations; however, this entire text should be consulted concerning the first capture theory of OSS.)

9. The entire subject of Venus and the interpretation of atmospheric data have been inordinately clouded by NASA's dominance of the press in the U. S. In light of the present paper, Venus is without doubt a young, highly active planet. NASA space scientists have insisted on creating, after-the-fact, *ad hoc* theories to explain almost all of the data (*i.e.,* the greenhouse effect, the "exaggerated" greenhouse effect, the colliding asteroid theory, etc.). The fact is that the Russians have gathered much data on Venus and had recognized many years ago the volcanic nature of Venus. (NASA only recently announced similar results.)

Historically, NASA scientists have pooh-poohed the Russian scientists and belittled them with the arrogance commonly found in U. S. astrophysics-astronomy circles (for an example, see OMNI (January, 1980), p. 38, "Disappearing Mountains"), thus hindering dialogue between the two groups. Note that some scientists disagree that the greenhouse effect is the sole cause of the high *surface* temperature of Venus (L. Greenberg, KRONOS IV:4; and Dr. Suomi, *Science News,* Nov. 3, 1979, p. 308). This entire issue is still under debate.

As recent as 1983, investigations of atmospheric aftereffects of the eruption of the Mexican volcano El Chichon by satellites, atmospheric probes, and Earth ground stations, have confirmed a number of suspicions of the author (see "The Atmospheric Effects of El Chichon", *Scientific American,* Vol. 250, Jan. 1984). Extensive aerosol clouds of sulfur based gases circled the globe for months after the three-day eruption. Detailed measurements showed a conclusive *cooling effect* as the cloud moved westward. This is in diametric opposition to the expectations of the greenhouse effect theory which would claim an *increase* in temperature due to trapped infrared radiation. The article concludes that: "Devine and Sigurdsson have found a good correlation between their estimates of the amount of sulfur gases volcanoes have released and decreases in mean hemispheric temperature." The authors, Rampino and Self, found the same result for El Chichon.

The sulfuric acid and other volatiles found in Venus' huge atmosphere are exactly what one would expect according to the present theory. (Note: if an asteroid collided with Venus with enough energy to drive off its oceans, as proposed by some space scientists, then how could Venus retain its atmosphere, complete with such gases as carbon dioxide and argon-36?) The planet's surface is clearly the result of recent volcanic activity, with some of the major volcanoes being still active.

10. A. Dauvillier, *Cosmic Dust,* p. 66.

11. W. M. Smart, *Celestial Mechanics* (N.Y., 1953), pp. 232-46. The paper by Forshufvud (KRONOS VII:2) proposes a nebular cloud, now dissipated, that he speculates may have originated from Saturn and claims that it may have aided in circularizing the orbit of the Venus Comet.

12. 1980-83 must be remembered in astronomical history as the years of the discovery of rings. Rings of matter have been discovered around numerous stars and stellar objects, while previous knowledge of the zodiacal disk and planetary nebulae (faint doughnut shaped rings long known to circle certain stars) has not received attention in the popular media.

Newly discovered rings are found in two forms: the first is the doughnut shaped nebular clouds found around Jupiter, Saturn, and as many as 50 nearby stars; the second is the Saturn type ring. The flat Saturn type rings have been found around Jupiter and Uranus in addition to new rings discovered around Saturn. The Sun is now known to have similar rings between 900,000 and 1,500,000 miles above its surface (see *DISCOVER,* (December, 1983), p. 14) and also between Mars and Jupiter (see *Minneapolis Star* (November 10, 1983), p. 3A) are found three stable rings. This paper contends that these rings are a normal attribute of stellar-like objects and that these rings constitute the sources for comet tail material during the discharge of the solar capacitor.

13. On the subject of precession, crustal shifting, and migration of the poles of the spinning planetary mantle, it is informative to spin a hard boiled egg and then spin a fresh egg. The hard boiled egg will continue to spin at the initial rate, whereas the fresh egg will quickly reduce its rotational rate since the interior is not bound to the shell and remained almost stationary during the initial spin, although the shell could initially spin. Viscosity quickly brings the shell and fluid interior into an equilibrium rotational rate. The Earth is similar to the fresh egg. Its mantle may be moved considerably by an external torque, but it will quickly resume spinning at the original rate even though the original poles had migrated to a new location.

14. In 1981, Cornell University astrophysicists, using the Arecibo (Puerto Rico) antenna, measured rapid and unexpected fluctuations in electron densities in the Earth's ionosphere.

In 1981, the space shuttle Columbia was observed to have an aura-like glow surrounding it as it orbited Earth in the upper ionosphere. In light of the present paper, this unexpected aura can be explained as due to the shuttle passing into zones of varying electrical potential. The spacecraft will adopt the potential of the surrounding medium. Ions will impinge on the craft and fluoresce, being attracted by the electrical charge on the craft. (See also footnote 44 of Part I of this paper KRONOS IX:1, p. 34.)

15. See, *op. cit.,* p. 210+, Plate VI.

16. P. L. Lamy, *Astron. and Astrophys., 72* (1979), p. 54.

17. J. K. Cline, *Celes. Mech., 19* (1979), p. 405.

18. H. Alfvén and G. Arrhenius, *The Moon, 5* (1972), p. 230.

19. J. M. Bailey, *J. of Geophys. Res., 76* (1971), p. 7827

20. S. F. Singer, *The Moon, 5* (1972), p. 207.

21. D. U. Wise, *J. Geophys. Res., 74* (1969), p. 6034.

22. B. N. Middlehurst and G. Kuiper, *The Moon, Meteorites and Comets* (Chicago, 1963), pp. 569-71. (This reference discussed comet splitting and variable brightness of comets.)

23. R. A. Lyttleton, *Mysteries,* p. 121.

24. *E.g.,* the electrical discharge between the Sun and its nebular ion cloud which lies beyond Pluto.

25. R. Berry, *Astronomy, 9* (March, 1980), p. 18.

26. N. F. Ness, *et al., Science, 204* (1979), p. 982.

27. Recent announcements of Voyager I and II data have shown that Dione of Saturn and the rings themselves also exhibit similar discharges. (*E.g.*, Minneapolis *Star*, 8/30/81.)

28. B. A. Smith, *Science, 206* (1979), p. 946.

29. D. Halliday and R. Resnick, *Physics, Part II* (2nd ed.), (N.Y., 1965), p. 753. (Figure 34-2.) This figure shows a current-carrying wire passing perpendicularly through a table top on which iron filings had been sprinkled. The iron filings align in concentric circles around the wire indicating the lines of magnetic flux. Note that a clear space exists around the wire. This is *exactly* the form of the concentric rings observed on Callisto's surface with a "crater" in its center. Callisto has a second smaller ring pattern on its opposite side.

30. Note that there is very little hope of ever verifying the nebular collapse theory of OSS by observation. An observational test of the present theory may be provided within a few years by the 1985-6 comet missions of the USA, France, Russia, Japan, and the ESA.

31. A. Dauvillier, *Cosmic Dust*, pp. 23-30.

32. F. Whipple, *Sci. American, 242* (March 1980), p. 124.

33. Electric fields have been ignored completely in NASA space probes because scientists did not expect to find such fields (an attitude which persists today). Most unfortunately, neither NASA, nor the European Space Agency could be convinced of the need for electric field sensing probes on comet fly-by satellites. They plan on verifying the ice ball comet model, and seem uninterested in any other suggestions.

34. (See for example, Dermott, Szebehely, Bass, and Ovenden.) A good deal of work has also been done on verifying the so-called "Bodes Law" of planetary spacing. [Also see M. M. Nieto, C. J. Ransom, and I. Michelson on "Bodes Law" in *Pensée IVR* VIII (1974), pp. 5-7, 44, 45. – *LMG*] Investigators generally assume the nebular collapse hypothesis and try to show why the nebular cloud left planetary space as observed today. A review of their work is beyond the scope of this paper; however, a few references can be cited: A. E. Roy, *Orbital Motion* (2nd ed., 1982); G. W. Wetherill, *Sci. American, 244* No. 6 (June, 1981). Note that the Wetherill article discusses a computer simulation of planet formation from planetesimals assuming the nebular theory. Essentially, he restates Laplace's original nebular hypothesis and completely ignores its fatal flaws (discussed in footnote No. 8 of Part I of this paper, KRONOS IX:1). It is obvious that he has a starting point (a set of orbiting protoplanets) and an end point (a solar system similar to ours) and at many points during the simulation, he "helps it along" by introducing "simplifying assumptions". He ignores many facts of celestial mechanics such as the inability of bodies to collide which are in similar orbits such as Saturn's two dancing moons. (Extensive computer calculations by S. F. Dermott of Cornell have shown that, in fact, such bodies will *never* collide, but will *always* dance around one another.) The reference to Safranov's 1961 simulations has been superseded by Szebehely's 1974 computer analysis which shows that, rather than coalescing, N-body systems that are given random initial conditions will eject the smallest bodies of the system, leaving only 2 bodies to orbit one another. He doesn't explain why the asteroid belt never formed a planet nor does he address the angular momentum problem of the Sun (why was the spinning, contracting Sun left with so little angular momentum?). Other completely ignored problems are the orientation of spin axes and spin rates of the planets, retrograde moons, the source of short and long lived radioactive elements in planetary cores, and the magnetic fields. But cf. the paper by E. Everhart, "Close Encounters of Comets and Planets", *Astron. J., 74* (June, 1969), pp. 735-750.

35. This is made possible because the system will constantly work to pull itself into the plane of total angular momentum, *e.g.*, Jupiter pulls down on bodies above its orbital plane and up on those below until all lie in the same plane at time equal to infinity. Statistical data on comets reflect this. Over 75% of comets orbit the Sun within 35° of the ecliptic, with over 50% within 10°, the majority moving in prograde orbits.

36. R. A. Kerr, *Science, 206* (1979), p. 40.

37. T. J. J. See, *Researches*, p. 193.

38. R. A. Lyttleton, *The Moon and the Planets, 23* (1980), p. 35.

39. "Disappearing Mountains", OMNI (January, 1980), p. 35.

40. F. Hoyle, *Highlights in Astronomy* (San Francisco, 1975), p. 46.

41. B. Mason, *The Lunar Rocks* (N.Y., 1970).

42. D. O. Froude, *et al., Nature, 304* (1983), pp. 616-618.

43. "Oldest Known Rocks Found in Australia", *Science News, 123* (1983), p. 389.

44. Very high energy particles will be found impinging on the comet nucleus during the discharge of the solar capacitor in highly active comets.

45. See also G. Talbott, KRONOS IV:2, pp. 3-25.

46. The following star-like characteristics attest to the fact that both Jupiter and Saturn are small stellar objects, with Saturn being much more active than Jupiter. This is a result of ongoing fusion triggered by energetic lightning bolts in their atmospheres.

    Rings: All three have stable flat rings (see footnote 12 above). The Sun has at least four such rings and its zodiacal disk; Jupiter has only one thin ring, whereas Saturn exhibits an extensive ring system with energetic electrical discharges between the rings and from the rings to the planet (observed as auroral belts).

    Interplanetary electrical discharges: the Io to Jupiter current sheet of 5 million amps has been detected; Dione of Saturn (as well as Saturn's rings) maintains a similar discharge, and comets which approach the Sun within the zodiacal disk form visible sunward electrical discharges (the sunward spike).

    Energetic particles: Surrounding Saturn is a plasma with particle energies higher than those found in the interior of the Sun. No known mechanism other than fusion could produce particles with such energies.

    Differential rotation: Saturn's and Jupiter's high velocity equatorial wind belts correspond to the differential rotation observed in the Sun's atmosphere, the wind speeds in Saturn being over three times those observed at Jupiter.

    High temperatures: Pioneer-Saturn measured the temperatures in Saturn's ionosphere to be 1250°K. "This high temperature requires an extensive energy source other than the Sun. This phenomena was also observed at Jupiter." (Quote from 1980 NASA news release *Science Highlights*.) See also M. Shimizu's paper in *The Moon and the Planets, 22* (June, 1980).

    Exothermic radiation: Both Jupiter and Saturn appear very bright at certain wavelengths, radiating two to three times the energy they receive from the Sun. Saturn emits more excess radiation for its size than Jupiter.

    Atmospheric conditions and lightning: The atmospheric circulation patterns of stellar objects are similar, with the violent storms producing energetic lightning bolts detectable as radio noise. The lightning bolts in Saturn's rings and in the atmospheres of Jupiter and Saturn are estimated to be millions of times more powerful than lightning bolts on Earth (see cover photo this issue).

    Footnote No. 41 of Part I (KRONOS IX:1) further discusses electrical phenomena at Saturn.

47. R. A. Kerr, *Science, 207* (1980), p. 292.

48. Astro News, *Astronomy, 9* (March, 1981), p. 58.

49. W. Sullivan, New York *Times* (July 1, 1980), p. C1.

50. "Jupiter and Saturn", *Sci. American* (December, 1981), pp. 90-108 (A. P. Ingersoll).

51. "The Atmosphere of Venus", *Sci. American* (July, 1981), p. 66 (G. Schubert & C. Covey).

52. R. A. Kerr, *Science, 209* (1980), p. 1219.

53. R. A. Kerr, *Science, 207* (1980), p. 289.

54. R. Berry, *Astronomy, 9* (March, 1980), p. 18.

55. B. Smith, *Science, 204* (1979), p. 955.

56. M. L. Kaiser, *Science, 209* (1980), p. 1238.

57. M. M. Waldrop, *Science, 210* (1980), p. 1107.

58. For a complete list of all radioactive isotopes and the energies of formation from other elements, refer to a *Handbook of Chemistry and Physics* (look in the index

under "Radioactive Elements"). Note that energies of particles necessary to initiate nuclear reactions in atomic nuclei depend more on the Coulomb barrier of the atomic nucleus rather than specific properties of the nuclei. At particle energies in excess of 20 MeV, however, a great variety of nuclear transmutations will occur.

59. S. E. DeForest, *J. Geophys. Res*, *77* (1972), p. 651.

60. In short, the induced electric dipole red-shift concept amounts to the following: A non-uniform electric field exists between a central star undergoing fusion (quasar, or other star-like object) and a surrounding nebular ion cloud (held back by proton wind pressure) as discussed in Parts I and II of this paper.

    Secondly, a gamma ray photon may divide in the strong non-uniform electric field of an atomic nucleus during electron-positron pair production. The photon energy is converted into the rest mass of the particles, but due to conservation of charge, the positive and negative charges must have existed in the photon prior to pair production. It is the induced electric dipole force that forces pair production in the gamma ray photon when near the atomic nucleus. By the same reasoning, all photons must contain charge pairs which will separate slightly when moving through a non-uniform electric field, causing an attractive (the induced electric dipole force is always attractive) force on the photon. Thus, photons leaving a central star's electric field (or passing by) will experience a slight central force which will reduce slightly the photon energy (or bend it towards the central star). This concept can be verified experimentally and will be presented in detail in Part III of this paper (in press).

* * *

APPENDIX I

A Sample Calculation For The Circularization Of An Orbit

A number of assumptions are made and a simple calculation is performed to illustrate the rapid orbital circularization that is possible due to the "tail drag" on a comet nucleus.

This appendix shows two results depending on the intensity of the electrical discharge between Sun and nebular ion-cloud, and on whether the discharge is "localized" or forms a complete breakdown of the solar capacitor. The first result provides an explanation for the clustering of comet orbits near the calculated position of the Oort cloud. The second result illustrates the rapid orbital circularization of comets which completely break down the solar capacitor.

Before presenting any formulas, a brief verbal explanation can be given: Laplace and Euler were early investigators of the effects of the "secular action of the resisting medium". Also, Encke suggested that such effects may cause comet wandering. These investigators showed that a resisting medium surrounding the Sun (becoming denser as one neared the Sun) would account for a spiralling sunward of any orbiting object and, secondly, would account for a reduction in the eccentricity of the orbit (the orbit became circularized). If the resisting medium were denser with increasing distance from the Sun, the inward spiralling would still occur; however, the eccentricity of orbit would *increase.*

After the discovery of the solar wind in the 1960s, scientists claimed that the resisting medium was an impossibility, since the wind of electrons, protons, and ions emanating from the Sun would quickly (within a matter of days) disperse such a cloud into the far reaches of the solar system. It was this discovery which firmly entrenched the ice ball comet model in the astrophysics journals. The assumption of an electrically neutral solar system also prevented scientists from realizing that there is, in fact, an excess current of positive charge in the solar wind (as presented in this paper).

Now it is known from observation that many rings of matter orbit the Sun, defying previously held theories, but also providing a source for comet tail matter as described in this paper. One ring lies very near the Sun while three others lie in the region between Mars and Jupiter. Possibly, other faint rings lie in the outer regions of the solar system. This paper also predicts that our Sun, like approximately 50 stars recently studied by the IRAS satellite, will have a doughnut shaped nebular cloud surrounding it past the orbit of Pluto. All of these rings are sources of comet tail material and, as mentioned, discharges may occur locally when comet nuclei enter these regions (similar to the discharges detected between the rings of Saturn) or a complete discharge (between the Sun and ring) may occur. The localized discharges have the effect of increasing comet orbital eccentricities whereas the complete discharge tends to decrease orbital eccentricity.

A) The "Oort effect": The nebular doughnut cloud estimated to lie past Pluto will be involved in localized discharges as comet nuclei enter this region of space around the Sun. The net effect will be to draw the comet orbital semi-major axis to the doughnut nebula. This is a direct result of Laplace's mathematical findings in his classic text *Mécanique Céleste.* We are not interested particularly in the time scales required to reach the final clustering of many comets around the doughnut nebula (such an analysis is far beyond the scope of this paper).

B) It is of greater importance to illustrate the possibility of rapid orbital circularization by the new comet theory, since this is paramount for those interested in Velikovsky's scenario of Venus becoming a planet from a comet, and having its orbit circularized in a very short time astronomically.

For this analysis, we assume: 1) the cometary discharge was complete (from Sun to nebular cloud or ring), 2) that the discharge became more intense while the comet nucleus was closer to the Sun, and became less intense at greater distances from the Sun, and 3) that the majority of mass of Venus was accumulated after its capture into the inner solar system by Jupiter (or Saturn).

These assumptions allow us to use the same basic formulas as used by Laplace and Encke. However, rather than the comet nucleus simply passing through a sparse resisting medium, large quantities of matter fall into the comet nucleus by powerful electrical forces as the comet discharges the solar capacitor.

The general results taken from Smart's Celestial Mechanics[11] are:

$$\Delta e = -\frac{4\pi ce}{h}$$

$$\Delta a = -\frac{4\pi ca}{h} \cdot \frac{1+e^2}{1-e^2}$$

$$\frac{c}{h} = 0.0007958 \quad \text{in this example}$$

where: Δe = change in eccentricity during one orbital period
Δa = change in semi-major axis during one orbital period
c and h are parameters related to the tail drag
e and a are the eccentricity and semi-major axis during one orbital period.

These results assume a resistive force of the form

$$R = \frac{cV}{r^2}$$

where: R = resistive force
v = velocity of comet nucleus relative to the source of tail material
r = Sun-comet nucleus separation

The following initial conditions are used, producing the final results in only 669 years as shown in the table below. Aphelion of the initial comet orbit lies near the orbit of Jupiter. The final aphelion lies at the present location of the planet Venus (encounters with Earth and Mars are not included since this example is only meant to illustrate the principles of orbital circularization.).

TIME (yrs)	APHELION (10^{12}m)	PERIHELION (10^{12}m)	e	location compared to present solar system
0	.74900	.68444	.0450	(comet near Jupiter)
570	.24306	.23584	.0151	(comet nears Mars)
635	.16243	.15918	.0101	(comet nears Earth)
669	.10873	.10727	.0068	(Venus' present orbit)

Due to the assumption of a constant rate of mass accumulation, (e.g., comet development depends on the solar discharge, not cometary mass) the majority of circularization occurs early in the 669 year simulation. Note that the increase in planetary radius in the final stages of evolution of this example is about one meter per hour.

APPENDIX II

Experiments Which Will Verify The Presently Proposed Theory For Comet Behavior

The four upcoming satellite fly-by missions (one to Comet Giacobini-Zenner and three to Halley's Comet) can verify the theory presented in this paper by noting a number of distinctive effects that will occur as the spacecraft approach the comets. These are predicted by the present theory and are totally inconsistent with any results of the ice ball comet model (unfortunately many space scientists have become adept at rationalizing almost any data to claim support for previously "accepted" theory by generating after-the-fact explanations. The author proposes at this time that the amount of after-the-fact explaining required by a theory be used as a measure of its incorrectness).

The U. S. probe (the International Sun-Earth Explorer 3) was diverted from environmental studies of the Sun in Earth's vicinity to an orbit that will pass through the tail of Comet Giacobini-Zenner in September, 1985. It carries an X-ray telescope, magnetometer, and particle detectors as well as a number of antennas. The other three probes of Japan, France-Russia, and the European Space Agency will all approach Halley's Comet in 1986.

A number of unusual results are predicted by the theory proposed in this paper and are not compatible with the ice ball comet model. They are as follows:

1) Tail material (positive ions and positively charged dust particles) will be detected by *direct* observation to move *towards* the comet nucleus. (Previously, Earth-based red shift measurements have been used to suggest that tail material moves away from the comet nucleus; Part III of this paper shows why these data have been misinterpreted.)

2) Spacecraft telemetry will begin to sense an acceleration towards the comet nucleus – far in excess of any expected gravitational field of a small "ice ball". (This is the result of the induced electric dipole force on the metal spacecraft.) This will also make it difficult to maintain spacecraft orientation while in the tail area.

3) This acceleration will cause an unexpected perturbation in the orbit of the satellite as it emerges from the comet tail. (In contrast, a negligible perturbation would be expected from a near-massless ice ball.)

4) If the satellites pass too close to the comet nucleus, they may be uncontrollably pulled into the comet nucleus for the same reasons.

5) The most conclusive evidence would come if one of the spacecraft would be diverted to pass between the comet nucleus and the Sun. The ice ball comet model would not anticipate any alteration in the solar wind in this area, since the ice ball is seen only to melt away in the tail direction. The present theory, however, would expect the following as the satellite approached a point between

the Sun and nucleus in the region of the comet nucleus. First, an intensified proton wind would be detected, followed by a substantial current of electrons (the electrons moving towards the comet nucleus) at the point between Sun and comet nucleus. This would be followed again by an intensified proton wind. This would substantiate the electrical discharge nature of comets as opposed to an ice ball melted by solar radiation.

6) Even if the spacecraft are not diverted into the region between the Sun and comet nucleus, the current sheets just mentioned should be detected as the spacecraft approach the comets.

7) X-ray activity may be detectable in the vicinity of the comet nucleus.

8) If the cometary discharges become intense enough in the comets being studied, both cosmic rays and X-rays (and possibly gamma rays) may be detected by the ISEE-3 satellite. None of these phenomena have been studied by Earth-based investigators since comets were believed to be melting ice. The author cannot overstress the complexity of reactions, both chemical and nuclear, that must occur in a highly active comet in which particle energies rise in excess of 20-30 MeV.

Note that both comet theories seem to explain data as observed from Earth, but it will be the effects listed above, which are only detectable by a spacecraft in the vicinity of the comet nucleus, which will establish the correctness of one theory or the other.

THE NATURE AND ORIGIN OF COMETS AND THE EVOLUTION OF CELESTIAL BODIES

(PART III)

J. M. McCanney

PREFACE TO PART III

Parts I and II of this paper (KRONOS IX:1, Fall 1983 & KRONOS IX:3, Summer 1984) introduced new concepts describing cometary behavior and solar system evolution. A number of basic theoretical results followed from the realization that there is an excess current of positive charge in the solar wind which emanates from the Sun. These included:

1) the formation of a "stellar capacitor" around stellar objects undergoing nuclear fusion in their atmospheres (the negatively charged stellar object is surrounded by rings and a doughnut shaped nebular cloud of ionized dust, molecules, and ions, forming an electrical capacitor which may discharge under certain conditions).

2) comets are asteroidal bodies (not ice balls) which discharge this stellar capacitor, developing a net negative electrical charge. The comet nucleus attracts quantities of dust and ions, forming the visible comet tail. The physics of these processes and the myriad of observed cometary phenomena are explained in Part I of this paper.

3) comet nuclei are captured by the solar system at random time intervals and evolve into the planets, moons and asteroids. Comets are accumulating matter and are not melting away as suggested by the ice ball comet model (IBCM). Only rarely do comet nuclei reach planetary dimensions. The members of the solar system have varying ages.

4) gravitational encounters with members of the solar system and the "tail drag" are the primary effects which move newly captured comets into stable non-overlapping orbits. The solar system is a dynamic ever evolving system.

5) the gas giants (Jupiter, Saturn and possibly Uranus and Neptune) maintain fusion in their atmospheres. The fusion is ignited by energetic lightning bolts in their turbulent atmospheres, implying that the Sun does the same. Observations of Jupiter and Saturn by Voyagers I and II show the same electrical

phenomena as observed around the Sun. Jupiter and the Sun were the original twin stars of our solar system, other bodies being captured one by one at later dates.*

6) gravity is not the sole force governing the cosmos. Electrical effects generally produce only subtle effects, but occasionally they may dominate the workings of the solar system.

7) major Earth altering events (caused by gravitational and electrical effects) may occur when large comets pass nearby.

Part II (*APPENDIX II*) proposed numerous experimental results of upcoming comet fly-by missions which will prove or disprove the concepts of this paper. The author would wish, as opposed to previous practices, that NASA and other space agencies make public all raw data, and not simply their interpretations of selected data.

The present paper (Part III) discusses two final concepts which follow from the presence of a solar capacitor. The first proposes a new source of red-shift in photons leaving a central star (the "Induced Electric Dipole Red-Shift" – IEDRS). Numerous in-lab experiments exist which can prove its validity. The other is the source of magnetic dynamos in stars and planets (and also magnetic reversals). Discussed briefly are biological evolution and the "Collective Fear of Velikovsky" that has permeated the scientific community for the past 35 years.

V. THE INDUCED ELECTRIC DIPOLE RED-SHIFT (IEDRS)

The search for an alternative cause of red-shifted photons in stellar spectral lines is not new. Many investigators have noted[(1)] that certain red-shift data are not explained by Special or General Relativity and some have attempted to search for new gravitational effects, none of which have successfully accounted for available data.[(2)]

The Hubble constant, which is used in conjunction with the special relativistic red-shift as a measure of distances to luminous stellar objects, has been questioned by many. It has been deeply engrained in the papers and interpretations of data found in the astrophysics journals for over a half century. Unfortunately, this is one of the primary reasons for its continued acceptance and the reluctance by some scientists to ponder its possible incorrectness. The Hubble constant is unlike other physical constants. It cannot be measured under laboratory conditions, but depends on a cyclic argument (red-shifted stellar spectral lines predominated, which some interpreted as indicating an expanding universe. As the expanding universe concept grew in acceptance, this led to further acceptance of the Hubble constant's use).

*Cf. H. Tresman and B. O'Gheogan, "The Primordial Light?", *SISR* II:2 (Dec. 1977), pp. 35ff.; I.Velikovsky, "On Saturn and the Flood", KRONOS V:1 (Fall 1979), p. 3. – *LMG*

It is proposed here that small induced dipole forces acting on photons can account for both the red-shift and bending of light around stellar objects (celestial bodies with ongoing fusion in their atmospheres). This follows from known observations of macroscopic and subatomic phenomena.

In atomic physics, gamma rays are known to split (electron-positron pair production) while in the intense central electric fields of atomic nuclei and charged subatomic particles. The photon energy is converted to mass and kinetic energy, but due to the law of conservation of charge, the (-,+) charge pair must have been contained within the photon prior to pair production. It must be the induced electric dipole force that forces the charge pair to separate.

Less energetic photons do not have sufficient energy for pair production (1.02 MeV is the minimum energy required), but they must similarly contain a (-,+) charge pair. This pair will separate slightly while in any non-uniform electric field, causing an attractive force (the induced electric dipole force is always attractive). This small force acting on photons as they travel over astronomical distances will reduce photon energy (the red-shift) and cause a bending of light in photons passing by a stellar object.

The IEDRS concept can explain the anomalous red-shift of photons observed at the Sun's edge and resolves two other mysteries.[(3)] One is the differing red-shifts of galaxies and certain quasars which appear to be associated visually with the nearby galaxies (suggesting that the red-shift cannot be used as a measure of astronomical distance).[(4)]

The second is the existence of planetary nebulae (ring clouds long known to circle certain stars). According to traditional astronomy,[(5)] these appear to be many light years from the central star because the red-shift is interpreted to give a great distance to the star from Earth. But if the red-shift is due to the IEDRS, then these stars may be much closer than previously thought, and the nebular clouds much closer to the central stars. Interestingly enough, the IRAS satellite recently found over 50 nearby stars with similar surrounding clouds, and this paper suggests that this is a common property of all stars (including the Sun). It is apparent from present knowledge of Pioneer 10 data, that the Sun's nebular cloud is far beyond the orbit of Pluto.

Another conclusion is that quasars are not superluminous objects at the edge of the universe, but are much closer than proposed by the use of the Hubble constant. With the previous paper's results,[(6)] it is apparent that quasars are the initial formative stages of normal galaxies (see footnote No. 39 of Part 1, KRONOS IX:1).

This implies that the bizarre theoretical extrapolations of General Relativity and the Big Bang Theory, such as "black holes", non-Euclidean space, and the "expanding universe" are no more than elements of some scientists' imaginations. Einstein himself was openly critical of General Relativity as "it did not include the total field (electric and magnetic)".[7] It is apparent that this was related to his interest in Velikovsky's work, although neither lived long enough to resolve the issue. Einstein's life-long search for a field unification theory also showed that he realized the importance of substantiating the famous "elevator analogy" with a rigorous theory. To date, no progress has been made by theorists to unify gravity with the electromagnetic equations of Maxwell; and the Principal of Equivalence remains without theoretical basis.[8]

Highly charged comet nuclei will also exhibit a red-shift in spectral lines regardless of the position or velocity with respect to the observer. Red-shift data exist for comets, but have always been interpreted as giving the velocities of tail ions as scientists have assumed that tail ions move *away* from the comet nucleus. "Tremendous tail accelerations" have posed an unsolved dilemma in interpreting this data. The IEDRS reconciles a new theory with this data, suggesting that the cometary red-shift is not a measure of tail ion velocities, but is a measure of electrical charge on the comet nucleus.[9]

The precession of Mercury must now be explained in a new context if General Relativity is incorrect. Note that other celestial bodies also "wander" from the orbits predicted by Newtonian Mechanics. The Moon has been known to wander since the later 1800's when exact data were first analyzed.[10] For this reason, exacting long term data on lunar wandering is needed for Io of Jupiter and Dione of Saturn (these are now known to interact electrically) and other smaller moons of the gas planets (i.e., moon 1979-S2 of Saturn) which have given evidence of being highly charged. Larmor's Theorem of Celestial Mechanics[11] provides for a slow precession of orbit for masses possessing charges. Additionally, Parts I and II of this paper suggest that mass accumulation (tail drag) will also affect the orbits of charged bodies.

VI. STELLAR AND PLANETARY MAGNETIC DYNAMOS

The standard explanation for planetary magnetic fields claims that electrical currents and resulting magnetic fields self generate in the interiors of stellar and planetary cores.[12] These "internal dynamo" theories, however, break down on two theoretical points: 1) they claim that purely mechanical processes produce electromagnetic effects, and 2) that fields can self-generate with no external mechanism to initialize or maintain this speculated process.

This paper proposes that the celestial magnetic dynamos are powered from outside by electrical interactions between charged orbiting bodies and both the stellar (or planetary) interior and upper atmosphere.

The empirical correlation between moons and planetary magnetic fields has been known for some time[(13)] and is unmistakably accurate, especially in light of recent Voyager I and II data. The original con cept of Houben and Dermott provided for a gravitationally induced current flow caused by tidal action of the moon on the planet's fluid core.

With the realization that charged bodies exist in space, and that the Moon, Mercury and other celestial bodies wander from Keplerian orbits, the following generalizations can be stated. The basic assumption is that the magnetic fields are induced by the star or planet spinning inside a slightly charged orbiting body (note: comets generally exhibit side effects which suggest that they are highly charged compared to bodies in circular orbits. Therefore, the effects of a highly charged comet, passing by a star or planet, will be to greatly increase and/or alter the magnetic field of the star or planet. This is discussed also in Parts I and II of the paper).

In support of the externally generated dynamo theory, the Sun's magnetic field is controlled by Mercury and, considering the former's size in comparison to Saturn and Jupiter, is relatively small.[(14)] Also, the Sun rotates relatively slowly (again compared to Jupiter and Saturn).

Mercury rotates only once per orbital revolution and has no moon, so it has a negligible magnetic field.

Venus similarly has no moon, essentially no rotational spin, and no magnetic field.

Earth, however, has a large magnetic field for its size as it spins on its axis every 24 hours inside a large moon. Furthermore, the Moon's orbit lies outside Earth's protective radiation belts (the Van Allen Belts) and is exposed directly to the solar wind. This implies that it will charge, since its orbit takes it to varying positions within the solar capacitor.

The Moon rotates on its axis only once every 27 days, and thus it has no magnetic field.

Mars rotates rapidly as does Earth, but only has two insignificant asteroidal moons and therefore has a negligible magnetic field.

There may be certain asteroids which possess strong permanent magnetic fields, as they must have cooled past the curie temperatures of constituent materials while in strong external magnetic fields.

Jupiter has a large magnetic field as it spins once in approximately 10 hours inside the orbit of electrically charged Io.

Saturn's field is not nearly as strong as that of Jupiter even though its spin rate is approximately equal to that of Jupiter. This is because Dione is responsible for Saturn's field. Dione does not exhibit the same level of electrical discharge to Saturn as Io does to Jupiter. Saturn's field is well aligned with its spin axis because Dione's orbit lies exactly in Saturn's equatorial plane.

Using this information, the properties of the magnetic fields of Uranus and Neptune may be anticipated. Uranus, even if it exhibits high levels of fusion in its atmosphere, will have a small field, probably well aligned with its spin axis.(15)

The internal dynamo theories, however, have poor empirical correlation. For example, Venus and Io have very active interiors, yet have no magnetic fields compared to Earth which is relatively cold and inactive, but which has a large magnetic field.

The externally generated dynamo proposal is enhanced when biological fossil records and magnetic reversal data are seen to coincide. The fossil record shows that a significant percentage of plant and animal species have died out concurrently with magnetic reversals, and that this is accompanied by periods of extensive volcanic activity, rapid stratification in lake beds, and crustal fracturing of the Earth's mantle. An excellent compilation of related references is available.(16) Any theory must account for all of these factors, which have occurred simultaneously on several occasions in Earth's history. Many of the theories proposed by scientists who have recently "discovered" celestial catastrophism only account for a subset of these factors.

Also, the recent highly acclaimed paper by Kopper and Papamarinopoulos(17) found a statistically significant correlation between human evolutionary changes and magnetic reversals.

With the proposed Earth altering event of 65 million years ago that ended the reign of the dinosaurs, there was associated an immense shower of meteoric material. This associates the event with an asteroid or comet. The associated magnetic reversals which occur in conjunction with rapid evolutionary changes indicate that this was *not* due to an asteroid colliding with Earth as suggested by Alvarez and others.(18) Although they have claimed that an asteroidal collision would force a geomagnetic reversal, no mechanism has been proposed which links a mechanical collision with electro-magnetic field generation.[19]

This paper suggests that the close passage of a highly charged comet with an associated meteor stream can explain both the worldwide presence of iridium in soil layers and the associated geomagnetic reversal. A simple test of the externally generated dynamo theory is to observe either solar (or other) magnetic field disturbances by comets passing close to the Sun. A passing ice ball would not be expected to affect a celestial magnetic field using the internal dynamo theories. Such effects would confirm both the electrical nature of comets and the externally generated magnetic dynamo concept. The Galileo Orbiter of Jupiter will be expected to detect such effects around Jupiter.

The MAGSAT satellite, which in 1981 ended an eight month examination of Earth's magnetic field, showed that the field had decreased by a small amount during that time. MAGSAT scientists extrapolated this trend[20] and predicted a geomagnetic reversal in 1200 years. It appears, however, that this is only the result of variable lunar charging in the solar wind.

The controversy in evolutionary biology[21,22,23,24] revolves around the interpretation of the fossil record. Either the data are interpreted literally (that short periods of devastation and genetic change preceded and followed long periods of genetic stability . . . the catastrophist argument), or, they are claimed to suggest that gaps in the fossil record exist because of lost data (the missing link hypothesis).

In genetic engineering, the genetic chemistry must be altered by external agents, as these molecular structures are inherently very stable over long periods of time compared to the life span of an individual member of the species. So, when a high percentage (i.e., 75%) of all species on both land and in the sea perish at one time, accompanied by a geomagnetic reversal, an influx of meteoric material, renewed volcanic activity, mountain building, and genetic change in the remaining species, most scientists agree that the Earth did not do this by itself.

The author agrees with Alvarez, Whipple, Clube, Napier, and others in that asteroids must have collided with Earth and caused extensive damage. This paper further suggests that many of the Earth altering events, including the 20 short epochs of mountain building[25] are attributable to a common cause . . . the occasional chance encounter between Earth and a comet which has been perturbed into the inner solar system. This adds to the colliding asteroid hypothesis, an "action at a distance" alteration of Earth. This paper differs from the colliding asteroid theories by claiming that comets are *not* ice balls, but are electrically charged asteroidal bodies which may rarely reach planetary dimensions. These few large comets may interact with the planets by gravitation, electrical discharge, and magnetic coupling (see also footnote 7, p. 76 of Part II, KRONOS IX:3).

This shows nature's plan of radical change through violence as with floods, forest fires, etc., and must be her way of assuring the continual mixing of the elements by random events allowing life to flourish and change.

VII. THE COLLECTIVE FEAR OF VELIKOVSKY

Unfortunately, the development of theoretical explanations of space probe data in certain circles has become a game of "keep Velikovsky from being right" and "preserve the accepted theories at all costs". An amazingly unscientific letter appeared in *Science* magazine following the Pioneer-Venus landing mission.[(26)] Editor Richard Kerr stated that, in spite of the fact that Pioneer-Venus data is far from explained by present theory, "few converts to the fringe are likely" and "the number of converts is probably the best measure of the power of the catastrophists' arguments". (By Kerr's own standards, catastrophism has found many "converts to the fringe".) This implies that the correctness of theory depends on a popular vote and not the objective analysis of data. It also warns astronomers that they will be singled out and ridiculed by strong peer pressures if they mention support for Velikovsky.

There is a high degree of correlation between Velikovsky's writings and the comet theory presented in Parts I, II, and III of this paper. The correlation is even more striking when one realizes that Velikovsky wrote his books before 1955 with only pre-1955 knowledge of what a comet really was. If he had written them after later discoveries regarding the true nature of comets, some might say that he was biased in interpreting his historical and geological findings. But this is not the case.

Quite recently, moreover, Newell[(27)] has published a review of data from Venus space probes which indicates that Venus may be at least 3 billion years younger than Earth. He also mentions, as have many other scientists, that the much heralded greenhouse effect cannot account for the molten planetary surface and that it contradicts the 2nd law of thermodynamics (heat cannot be pumped from the cooler clouds to the warmer planetary surface).

Theories in other fields such as plate tectonics, Darwinian evolution (gradualism), etc., have followed the lead of astronomical theory which has claimed that the solar system formed 4.5 billion years ago and that the order of the inner planets has since remained unchanged. What we are now seeing is that the wall that was built when "scientists confronted Velikovsky" a decade ago is crumbling badly. Scientists can no longer fall back on their view that "Velikovsky has been proven wrong".

The theory proposed in this paper is based on sound physics and the most recent space probe data. It now supports the contention that Venus may have been "born of Jupiter" and that the ancients did indeed witness the fetus of birth.[28] Venus, however, was not projected from Jupiter's interior as Velikovsky speculated, but was *captured* and tossed into the inner solar system by a mechanism commonly known in Celestial Mechanics. Velikovsky's realization of historic descriptions of this event will someday be recognized as one of the great discoveries of the 20th century.

From glaciation data and the fossil and faunal records, it appears that the pre-Venus north pole of Earth was somewhere in the middle of the Canadian Northwest Territories as the "glaciers" of the last "ice age" never extended into Siberia, which was apparently more temperate at the time. (See also footnote 13 of Part II, KRONOS IX:3.)

The "red hand of death" spoken of in ancient legends may have been auroras caused by the current sheet between the comet Venus and Earth, just as Io's current sheet causes visible auroras as it passes over Jupiter's dark side; or, it may have been the visual effects of flaming hydrocarbons (oils and tars) as they rained into Earth's atmosphere from Venus' huge cometary tail.

The columns of smoke extending into the clouds that sounded as tremendous drumming[29] occurred when the comet to Earth discharge became a single beam snaking between comet and Earth (see cover photo and photos 1 to 6 (pp. 64 and 69), KRONOS IX:3). Numerous other historically observed phenomena have been correlated to the present theory, and will be the topics of further papers.

Questions still remain concerning a recent birth for Venus. But, there is mounting evidence suggesting that Venus is indeed a youthful planet; and it could very well have been the comet of the ancients that Velikovsky identified and discussed in *Worlds in Collision.*

Since *Scientists Confront Velikovsky*[30] and the 1974 AAAS Velikovsky trial, more cosmological data has been gathered than had been accumulated in the previous 300 years. A catastrophic band wagon has been formed in astronomy, geology, evolutionary biology, and related fields, but with no credit given to Velikovsky. Has data and observation pointed towards an electrical universe, epochs of celestial catastrophes affecting Earth, etc.? Let the reader decide.

Velikovsky's main difficulty was that of a man far ahead of his time in a world of academic elitism. He disagreed with the "experts" who used their resources to label him a crackpot. In discrediting Velikovsky, Sagan, *et al.* simply restated "traditional" theories and it became apparent that catastrophism did not fit. As stated by Carl Sagan: ". . . of the ten tests of Velikovsky's work . . ., there is not one case where his ideas are . . . consistent with simple physical theory and observation."[(31)]

The present paper illustrates why Venus the comet was not consistent with the physical theories of 1974. Venus the comet *is* compatible with physical theory in 1984 and, furthermore, with general trends in all associated fields. Time will tell whether data from upcoming comet probes will support the ice ball comet model (IBCM) or the present theory, and whether the data will be interpreted by NASA and related groups only in light of the IBCM. Fortunately, numerous other international agencies will also have comet probes, and they do not fall under the spell of NASA's protectionism of data and theory.

CONCLUSION TO PART III

A new self-consistent theory of comet behavior and solar system evolution has been put forth. Recent data from space probes have been correlated to the theory and further experiments have been defined to prove or disprove the competing comet theories. Many of these experiments are Earth-based, but as space probes have repeatedly shown, there is no substitute for close range data.

REFERENCES AND NOTES

1. H. C. Arp, *Astrophys. Letters,* 9 (1971), p. 1; and J. W. Sulentic, *Astrophys. J. (Letters),* 265 (1983), pp. L49-L53; and H. C. Arp, private communication.
2. A little understood fact is that Einstein's General Relativity was a political football in the astronomical community and was railroaded into prominence to maintain the belief that gravity could explain all observations in the cosmos. [See H. C. Dudley, "The Personal Tragedy of Albert Einstein," KRONOS I:4 (Winter 1976), pp. 55-67. – *HAH*] Einstein was never satisfied because electromagnetic fields were not included. I believe that the leap of intuition found in the principle of equivalence (i.e., the famous elevator example) is incorrect. Einstein spent the last years of his life looking for the unification of gravitational and electromagnetic fields (the long sought unification of fields). This unification, if found, would necessarily result in equations relating gravity to Maxwell's four equations. No such relationship has ever been found (either experimentally or theoretically). (See *DISCOVER* magazine, December, 1983, pp. 44-53.) In this respect, the necessary cornerstone of General Relativity has never been found. This was why Einstein was so interested in the electrical findings of Velikovsky. I believe that more data can be explained by the IEDRS concept than by General Relativity or the Hubble constant.

   1982 saw the emergence of scientists questioning the validity of Einstein's calculations concerning Mercury. Although a tremendous amount of theoretical work has been done on General Relativity (and *over* publicized), very little *experimental* work has been successfully performed in the past 50 years. The greatest topic of discussion among experimentalists concerns the correction factors that make the data fit theory. In most cases, data are simply irreproducible and remain unpublished. Unlike Special Relativity, General Relativity lacks experimental verification.

In 1981, I attended a colloquium at Cornell University which dealt with scientists tracking radio signals from satellites in well known orbits around the Sun (to measure the bending of light rays around the Sun). Their data showed only irreproducible results. The meeting turned into a discussion of what "correction" factors could be added to the data to make them fit theory. Unfortunately, irreproducible data is never published.

Dr. H. Arp's findings (that some quasars are visually associated with galaxies that exhibit substantially different red-shifts) has cast doubt on the validity of the use of Hubble's constant as a measure of astronomical distance. Using the induced electric dipole red-shift concept, the large red-shifts found in quasars can be explained as due to enormous proton winds emanating from the quasar which in turn causes large red-shifts in photons leaving the quasar. Quasars are not at the edge of the universe nor are they receding at near the speed of light.

3. M. W. Browne, "Cosmic Yardstick: Is the Yardstick in Error?", *New York Times* (Oct. 15, 1979), p. C1; and W. Sullivan, "Cosmic Bridges Suggest Quasars are Nearer Earth", *New York Times* (March 22, 1983), p. C1.
4. J. W. Sulentic, *op. cit.*, p. L49.
5. S. A. Kaplan, *The Physics of Stars* (N.Y., 1982), p. 149.
6. J. M. McCanney, "Continuing Galactic Formation," *Astrophys. Spa. Sci.*, 74 (1981), pp. 57-64.
7. A. Einstein, *Out of My Later Years* (N.Y., 1950), p. 48.
8. G. Taubes, *DISCOVER* (Dec., 1983), pp. 44-53.
9. The Venus Orbiter spacecraft will be repositioned by the NASA Ames Research Center team when Halley's comet is near perihelion. It will measure both light intensity and wavelength. From the information available at time of writing, it is not clear whether the comet will pass to the inside of Venus' orbit. If it does, this will provide a definitive test for both the electric comet concept and the IEDRS concept.

   If comets are ice balls with tail ions moving away from the Sun, then the Venus Orbiter should detect a *blue*-shift due to the doppler shift of light from *approaching* tail ions. If, however, the present paper is correct, then a red-shift will occur for two reasons, 1) the doppler shift of light from *receding* tail ions and 2) the induced electric dipole red-shift due to the charged comet nucleus. The correct position of the comet (i.e., between the Sun and observer) is crucial to the experiment. If an observer attempts to determine doppler shift with the comet on the opposite side as the Sun, the IEDRS will be misinterpreted as giving the velocities of receding tail ions as assumed by the ice ball comet model (such data already exists and has indeed been misinterpreted by astronomers).
10. T. J. J. See, *Researches on the Evolution of the Stellar Systems,* Vol. II (Lynn, Mass., 1910), pp. 274-92.
11. Symon, *Mechanics* (Reading, Mass., 1964), p. 283.
12. E. N. Parker, "Magnetic Fields in the Cosmos", *Scientific American* (August, 1983), pp. 44-55.
13. H. C. Houben, *"Tidal Dissipation in the Solar System and the Possibility of Tidally Driven Planetary Magnetic Dynamos"*, Cornell University thesis (Ithaca, 1978).
14. A subtlety here is that, although Mercury is the smallest planet (next to Pluto), it maintains the largest *total energy* of all the planets including that of Jupiter. This is because the equation for total energy in celestial mechanics for an orbiting body only depends on the inverse of the orbital radius. [Also cf. "The Sun's Magnetic Field," KRONOS II:3 (Feb. 1977), pp. 78-80. – *LMG*]
15. In October, 1982, Voyagers I and II data were released showing an electrical discharge flowing between Dione and Saturn, similar to the discharge between Io and Jupiter. The

same mechanism that electrically charges comet nuclei orbiting the Sun is at work around the giant gas planets (as they too have ongoing fusion in their atmospheres and proton wind supported capacitors). There are constant discharges of their capacitors by way of Dione and Io, resulting in net negative charges on these moons (as with Mercury and comets orbiting the Sun). These electrical effects drive the magnetic fields of Jupiter and Saturn. It will be expected that Uranus and Neptune will follow suit.

It appears that their magnetic fields depend on: 1) the quantity of solidified iron in their mantles, 2) the magnitude of charging of their moons and the levels of discharge from moon to planet, and 3) the magnitude of their stellar capacitors.

16. P. Warlow, *J. Phys. A: Math. Gen.*, 11 (1978), pp. 2107-2130; P. Warlow, *The Reversing Earth* (London, 1982).

17. J. S. Kopper, and S. Papamarinopoulos, *J. of Field Archaeology*, 5 (1978), p. 443.

18. It has been suggested by Alvarez and F. Whipple that a large asteroid collided with Earth and melted as it sprayed an iridium laden dust cloud to hang in Earth's atmosphere for years afterwards. (C. Sagan used similar assumptions to promote his world-wide winter following a nuclear war.) These have been criticized by many scientists, since it is now known that dust clouds from volcanoes do not remain in the atmosphere for more than a few days. Only molecular aerosol clouds remain to circle the globe. (See: C. Pellegrino, *Astronomy*, 9:4 (1981), p. 66 and, "The Atmospheric Effects of El Chichon," *Scientific American*, Vol. 250, Jan., 1984.)

19. Geomagnetic reversal due to a close gravitational encounter with a second celestial body has been suggested (See: *Velikovsky Reconsidered* (N.Y., 1977), p. 172 and P. Warlow, *J. Phys. A: Math. Gen.*, 11 (1978), p. 2107). The suggestions are that the Earth is "flipped" by precession of the rotating Earth in the gravitational field of a second body. This effect, however, has been greatly overestimated. Calculation shows that for an Earth-sized object stationed at two Earth radii from Earth, it would take at least 28 years to cause a 50° precession. This is far too long for a passing celestial body to remain in Earth's vicinity.

20. K. J. Rose, *OMNI* (March, 1981), p. 18.

21. See KRONOS VII:4 (Summer 1982) on "Evolution, Extinction, and Catastrophism".

22. J. Davis, *Astronomy*, 9:4 (1981), p. 6.

23. R. Lewin, *Science*, 210 (1980), p. 883.

24. J. O'Keefe, *Nature*, 285 (1980), p. 309. (Since 1981 the reference list of articles and books on the evolution debate has grown extensively.)

25. R. A. Lyttleton, *Moon and the Planets*, 18 (1980), p. 13.

26. R. A. Kerr, *Science*, 207 (1980), p. 293 (also read the previous pages of the "Research News" article on which Kerr is commenting).

27. R. E. Newell, *Speculations in Science and Technology*, 7:1 (1984), pp. 51-57.

28. Cardona has been making a case (see KRONOS VII:1, pp. 56-67, VII:2, pp. 29-40, VII:3, pp. 3-14, VIII:4, pp. 1-16) from historical data that Venus may have been a "child" of Saturn, not Jupiter. For our purposes, this makes no difference in terms of Celestial Mechanics, since all the gas planets are known to capture comets into the solar system, and each has associated "families" of comets. [Also see the article "Ejections, etc." by Cardona elsewhere in this issue. – *LMG*]

29. L. Ginzberg, *The Legends of the Jews*, Vol. II-III (Philadelphia, 1910).

30. C. Sagan, *et al.*, *Scientists Confront Velikovsky* (Ithaca, 1974).

31. *Ibid.*, p. 89.